THE VENOMOUS
AND
POISONOUS MARINE
INVERTEBRATES
OF THE INDIAN OCEAN

*To Jane
c̄ kindest regards
Jim. Russell
4/1997*

FINDLAY E. RUSSELL
RACHAKONDA NAGABHUSHANAM

OXFORD & IBH PUBLISHING CO. PVT. LTD.
New Delhi   Calcutta

This work is related to Department of Navy Grant N00014-85-J-1259 issued by the Office of Naval Research. The United States Government has a royalty-free license throughout the world in all copyrightable material contained therein.

ISBN 81-204-1013-0

Published by Mohan Primlani for Oxford & IBH Publishing Co. Pvt. Ltd., 66 Janpath, New Delhi 110 001. Typeset at Radiant Printers, New Delhi. Processed and printed at Baba Barkha Nath Printers. 26/7 Najafgarh Road Industrial Area, New Delhi 110 015.

1-N6-5.5

PREFACE

Anyone who writes a book on science is likely to put down as many errors as truths; science is merely a compendium of the facts of the period as they are known to be. The authors trust that they have not been presumptuous in setting down what they believe is an orderly and succinct review of our knowledge on the venomous and poisonous marine animals, putting particular emphasis on those creatures of the Indian Ocean as it encompasses India. Although the text is primarily designed for the biologist or zoologist, especially those at the undergraduate or graduate level, we have tried to include important data on the systemics and taxonomy of these marine animals, the structure of their venom or poison apparatus, the nature of the chemistry, and the physiopharmacology and immunology of their toxins. We have also attempted to provide a brief synopsis of the clinical significance of their poisonings. In so far as possible, we have endeavored to avoid abstraction and supposition.

The principal effort of this work has been to provide a practical text for those interested in the venomous and poisonous marine animals, be they student, teacher, researcher, public health worker, food scientist, fisherman, medic, or just the general reader. We have tried to furnish sufficient background and history upon which the interested person can develop a research program or lecture, or provide data of significance to the public. For the student and researcher, we feel it is particularly important not only to have a background familiarity with the current status of the scientific data but also a sufficient historical perspective upon which to direct research projects or teaching, or to investigate the development of useful drugs from marine creatures.

With these thoughts in mind we have arranged the text by phyla, describing those poisonous or venomous genera and species that are of primary interest with respect to their contact with mankind, or the toxicological importance of certain of their tissues. A brief consideration of the clinical significance of the poisoning that can be precipitated by these animals is also provided, not only for those working with these marine animals but also for public health workers or physicians who may be called upon to treat poisoned victims. In all of these matters, we apologize for any errors of omission or commission.

FINDLAY E. RUSSELL
RACHAKONDA NAGABHUSHANAM

CONTENTS

ACKNOWLEDGMENTS

Anyone who has studied marine toxinology, particularly its history, will appreciate the wealth of data B.W. Halstead has provided in his monumental *Poisonous and Venomous Marine Animals of the World* (1965–1970). The authors are deeply indebted to Dr. Halstead for permitting us to reproduce a number of photographs from that work, and for providing other information relevant to this text. His friendship over the past 40 years is greatly appreciated.

Some of the data are taken from a symposium on bioactive compounds held in Goa, India, in 1989.

The financial support of American Institute of Biological Sciences and the U.S. Office of Naval Research is gratefully acknowledged.

The authors also wish to thank the Council of Scientific and Industrial Research of India, the Marathawada University, and the University of Arizona. Grateful thanks to the following for their assistance with parts of the manuscript: D.W. Hessinger, D.A. Bakus, G.R. Southcott, R. Endean, K. Steidinger, N.C. Bucknall, and the late H.A. Reid among others.

FINDLAY E. RUSSELL
R. NAGABHUSHANAM

INTRODUCTION

This text is intended to present an overall review of the venomous and poisonous invertebrates of the marine waters about India north of the equator. Another interest is to examine the general chemical and pharmacological properties of the toxins of these animals and their significance in the animals themselves, as well as to the biota. A further contribution of the book is to present a practical approach to the first aid and medical management of the human poisonings caused by these creatures. Finally, in these days of advanced technology, the use of marine toxins (or their fractions) as tools in the study of biological membranes has become increasingly important and of significance to every student working in the fields of biology and medicine. We have attempted to explore this concern. Although the text has been oriented with the student and teacher in mind, it should provide the more advanced researcher, such as the electrophysiologist or investigative toxinologist, with sufficient data and history for undertaking more definitive research on these animals and their toxins. Hopefully, it will serve as a reference work for those interested in various drugs of the sea programs.

Historically, the venomous and poisonous marine animals have held a fascination for man since the beginnings of written record. Early peoples often attributed the consequences of their bites and stings, or their ingestion, to forces beyond nature, sometimes to vengeful deities thought to be embodied in the animals themselves. To these early peoples the effects of the bites or stings by animals were so surprising, so varied and often so violent that their injuries were shrouded with much myth and superstition. Even today, considerable folklore about these animals still exists. The task of separating fact from fiction is often a formidable one, and one not always lightened by the passage of time (Russell, 1980a).

In general, animals use their toxins in offence, as in the gaining and digestion of food or in defence, as in protecting themselves against predators. In some cases, a venom may serve both functions, as in the spitting cobra and, perhaps, in the scorpion. There may be other uses to which a venom is put. The "tagging" of prey by rattlesnakes (Chiszar *et al.*, 1982) may be a function of the animal's venom, although this has yet to be demonstrated.

In addition to the animals themselves, the study of the marine toxins, *marine toxinology*, has also enjoyed a fascinating history. The complexity of these substances, as well as the complicated and sometimes unknown means whereby they endanger or ⋅ destroy life, have invited speculation, exaggeration and even pure fantasy. During the past two decades, however, considerable progress has been made in our understanding of the chemistry and physiopharmacology of these marine toxins. This is demonstrated by the many hundreds of published articles on their properties (See Russell *et al.*, 1984).

On the basis of laboratory experiments or human encounters there are said to be more than 1200 marine animals known to be venomous or poisonous (Russell, 1984). These range from unicellular protistans to certain of the chordates, and while their numbers may sometimes be quite large they do not produce major ecological effects by virtue of their toxicity alone, nor are they other than a local danger to human health and economy. One can only guess at the number of still unknown marine species that may eventually be shown to contain toxic substances or be a danger to humans or marine organisms. With respect to the Arabian Sea and the Bay of Bengal, there are no valid data on the numbers of species of toxic creatures. Neither are there reliable public health statistics on the bites, stings or poisonings by marine creatures.

The term *venomous animal* is usually applied to those creatures capable of producing a poison in a highly developed secretory organ or group of cells and which can deliver this toxin during a biting or stinging act. *Poisonous animals* are generally regarded to be those whose tissues, either in part or in their entirety are toxic. In reality, all venomous animals are poisonous but not all poisonous animals are venomous. Animals in which a definite venom apparatus is present are sometimes called *phanerotoxic* (Gk. Ψανερόσ, evident + τοετχόν, poison), while animals whose body tissues are toxic are called *cryptotoxic* (Gk. Χρυπτόσ, hidden). The cobra, stingray and black widow spider are venomous or phanerotoxic animals, while the blister beetles, certain puffer fishes and toads are said to be poisonous or cryptotoxic (Russell, 1965a).

Although the terms *venomous* and *poisonous* are often used synonymously, more recently investigators have tried to confine the use of the term venomous animals to those creatures having a gland or group of highly specialized secretory cells, a venom duct (although this is not a constant finding) and a structure for delivering the venom. While there has been a tendency to employ the term *venom apparatus* to denote only the sting, spine, jaw, tooth or fang used by the animal to inject or deliver its venom, most biologists now use the term in its broader context, that is, to denote the gland, duct and the sting or fang. The associated musculature is sometimes included. Poisonous animals, as distinguished from venomous animals, have no such delivery apparatus; poisoning by these forms usually takes place through ingestion (Russell, 1965b).

Historically, the words *toxicity* and *lethality* have been frequently interchanged. Even today, some workers use the word toxicity when they mean lethality. Most toxinologists, however, try to differentiate between the two. In the opening address before the International Society on Toxinology in 1962, one of us (FER) indicated the more broad use of the word toxicity:

> A poison that causes a large ulcerating lesion or other tissue defect or a response of a deleterious nature is certainly toxic, while not necessarily lethal, and such pharmacological activities should not be overlooked. Toxicity might also include changes in the blood cells, coagulation defects, changes in nerve-muscle transmission, or any of a number of other

deleterious changes that do not necessarily end in death. Certainly, when one uses the word lethality he is being specific; that is, he is defining the end point as death, and only that end point. Such statements as "copperhead venom is not toxic" are unfortunate, for aside from the fact that the venom of *Agkistrodon contortrix* produces tissue changes, it is also lethal, even though the LD_{50} is much higher than for the venom for *Crotalus* species.

In describing a venom, therefore, it is important that its toxic and/or lethal properties be carefully defined. Even though some fractions of a marine venom may not be lethal, their effects may be of sufficient physiopharmacological or clinical importance to warrant specific definition.

A. History and Folklore

One of the first descriptions of a poisonous marine animal is found in the hieroglyphics of the tombs of Ti (Fifth Dynasty, *c.* 2700 B.C.), where the poisonous pufferfish, *Tetraodon stellatus*, is represented by the hieroglyphic inscription *crept*, *spt* or *shepet*. Its toxic properties appear to have been known even at that early period (Halstead, 1965). In about 1500 B.C., Moses records a possible toxic dinoflagellate bloom: ". . . and all the waters that were in the river turned to blood. And the fish that was in the river died; and the river stank, and the Egyptians could not drink of the water of the river" (Exodus 7:20–21).

In the *Chinese Materia Medica of Fish Drugs*, based on the writings of early Chinese workers (A.D. 618–1644), one can find a list of many venomous and poisonous fishes known to be dangerous to man (Read, 1939). Halstead (1965-1970) believes that the mention of the yellowtail or amberjack, *Seriola*, during the T'ang Dynasty (A.D. 618–907) as "a large poisonous fish fatally toxic to man" is the first description of a ciguateric fish. Pufferfish poisoning was known as early as the Sung Dynasty (A.D. 960). With respect to venomous fishes, the caudal spine of the stingray was used in the Orient as a spear head during the times of the Ainu, 2000 years ago, and perhaps even before that by the Jomon people (5000–2000 B.C.). The latter may have also used stingray poison as a powder or paste on their stingray spear heads (Tomlin, 1966; Bisset, 1976). Tokunai (1790) noted that the Ainu used the dried caudal spine of the Japansese stingray, *Dasyatis akajei*, with the integumentary sheath intact, as a weapon.

But the spine also had a more exotic use in sorcery during early times. If one wished to attract a particular charming girl, he put stingray spines in her footprints as she walked through the snow. This caused her to suffer a severe headache. Then he told her what he had done, whereupon she turned and appealed to him to rid her of her headache (Natori, 1972).

The Greek physician-poet Nicander (384-322 B.C.) notes the venomousness of the stingray and likens the moray eel to the viper because "its bite poisons with a venom like that of the viper." He wrote that some investigators believe that eels can leave the sea, driven by desire, to copulate with vipers. In his *Theriaca*, Nicander notes the weeverfish, *Trachinus*, as the sea dragon.

The most exhaustive and credulous early writer of natural and medical history fact and fiction, however, was Gaius Plinius Secundus (A.D. 23–79), whose voluminous *Historia Naturalis* contains numerous fascinating accounts of venomous and poisonous animals, including sections on the weeverfish and stringray. Describing the stingray, he writes:

> So venomous it is, that if it be sturchen into the root of a tree, it killeth it: it is able to pierce a good cuirace or jacke of buffe, or such like, as if it were an arrow shot or a dart launched: but besides the force and power that it hath that way answerable to iron and steel, the wound that it maketh, it is therewith poisoned.

So frequently quoted were Pliny's works that Hulme (1895) commented:

> Several writers of antiquity influenced the medieval authors, but it is scarcely necessary to detail their labours at any length, since if they lived before Pliny he borrowed from them, and if they lived afterward they borrowed from him, so that we practically in Pliny get the pith and cream of all.

As Klauber (1956) has so aptly concluded, "Pliny's *Historia Naturalis* was the funnel through which we can watch the

Figure 1. A daily life scene from the tomb of the Egyptian pharoh, Ti of the Fifth Dynasty (*ca.* 2750 B.C.). Among the fishes shown is the puffer *Tetrodon stellatus*, known to the ancient Egyptians to be poisonous (Keimer, 1955). (From Gaillard, 1923).

DE LA PASTENAQVE. CHAPIT. XXXI.

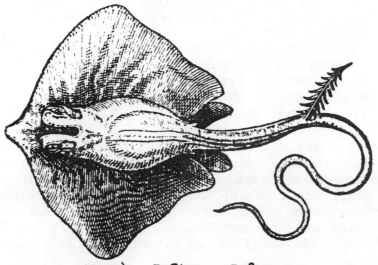

Τρυγών, *Paſtinaca*, *Paſtenaque*.

DE LA VIVE, OV DRAGON MARIN.
CHAPITRE XXXII.

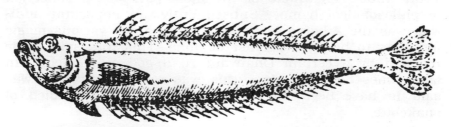

Δράκων θαλα΄σιος, *Draco marinus*, *Viue*.

Figure 2. The stingray *Dasyatis pastinaca* (L.) and the greater weever *Trachinus draco* (L.). (From Grevin. 1571).

ancient folklore pouring down into the medieval and modern worlds."

Most writings on venomous and poisonous marine animals following the Greek-Roman period tended to follow the descriptions of Nicander, Aristotle, Galenus and Pliny. During mediaeval times, Avicenna's (A.D. 980–1036) *Q'anum* contained a good description of venomous and poisonous animal bites and stings and their treatment, and Peter of Abanos' (A.D. *c.* 1250–1316) *De Venenis* detailed some entries on marine animal injuries. Although Gazio (1450–1530), Peter Martyr (1457–1526), Rondelet (1507–1557), and Gesner (1516–1565) described various dangerous marine animals in their writings, the most important work of the Renaissance was Jacques Grevin's *Deux livres des Venins* (1568). Grevin characterized the stingrays, weeverfish and sea hares, among other marine animals, and attempted to put the various creatures into some sort of categorical system. Grevin's work was followed by the contributions of Autenreith (1833), Bottard (1989), Phisalix (1922), and Pawlowsky (1927), among others.

In general, the history and folklore of Indian medicine is relatively easy to follow, since many of its practices are of ancient origin and have continued into modern times. The oldest and most respected Hindu works were known collectively as the Veda, the earliest portion being the *Rig Veda* (1500 B.C.) which for the most part was a collection of philosophical hymns. Subsequently, an important addition was the *Atharva Veda*, which contained various magic formulas to be used against demons. This was said to be received by a certain Dhanvantari, a god of medicine. In later times however, his status was reduced until he was said to have been an earthly king or one who died of snakebite.

During the Brahminical period (800 B.C. to A.D. 1000) the practice of medicine was almost completely controlled by the Brahmin priests and scholars. The body of medical literature that grew out of this period became known as the *Agur Veda*. In the the beginning, the Brahmins used three basic texts: the *Charaka-Aamkita*, written by Charaka *c.* A.D. 200), the Susruta-Aamkita, perhaps first written by Susruta around 50 B.C., and the *Vagbhata* by Vagbhata around A.D. 700. Of these three, the Susruta is considered the most remarkable and it soon became the basic book of Hindu medicine.

Because Hindus were prohibited by their religion from cutting the body after death, their knowledge of anatomy was very limited but their knowledge of disease was quite remarkable. Susruta gives 1,120 disease states, including wasting disease (tuberculosis), smallpox, fevers, etc. Their surgery was far advanced. Susruta also lists 760 plants of medicinal importance. Particular attention was paid to aphrodisiacs and poisons, as well as to antidotes for the bites of venomous snakes and other animals. The Hindus are believed to have borrowed some of their medicines from the Greeks, and some of the Greek drugs are known to be of Hindu origin. Ancient surgical procedures attained their highest development among the Hindus.

It is obvious from the numerousness of venomous snakes throughout India that Hindu mythology was represented by many snake divinities. It is also obvious that the myths and even the treatment of venomous animal injuries, which had their foundations in snake lore, trickled down to the spiders, scorpions, and even to the venomous marine animals (Russell, 1980a).

Indian serpent worship began along the lower Euphrates, among people of Turanian origin, and then spread with them to every part of the land wherein they settled (Fergusson, 1893). Among the best known of the Indian snake deities are the Nagas. They were a powerful race of serpent gods, sometimes assuming human form. Then there were snake kings, such as Takshaka, whose glittering capital was the glory of the under world. There was the serpent goddess Manasa, who even today is worshipped by some in Bengal, where her powers are often called upon against the ever-present danger of venomous snakes. Statues of divinized Nagas are still worshipped in southern India. These statues are usually found under a tree, often in an uncultivated area so that the god-snakes can have their own areas in which to meditate. It is believed that if a snake is given a sufficient area it is more likely to avoid human habitation, as well as to have more time for its divine activities. The Nagas and their wives, The Nagins, are sometimes described in various Indian tales of surprise and trickery. This is probably due to the great speed and reclusiveness of snakes. In epochs of cosmic rest the great snake Sesha formed a bed for the god Vishnu, and shielded him from the sun with his seven raised heads (Russell, 1980b).

In many parts of ancient India, snake worship was closely akin to ancestor worship. It was a Hindu familial cult, related to the protection of the home, its living and its dead. In some cultures the second self of a dead relative was thought to frequent his former house. How else would it be possible for the living in the house to see him in their dreams? Similar beliefs in returning spirits are, of course, common to every culture, the principal difference is that in certain Asian and Hindu myths the serpent often represents the soul of the deceased. Although the folklore of venomous marine animals is far less steeped in myth, the treatments of the injuries inflicted by these animals have obviously followed those for snakebite. Indeed, scarification, blood-letting and cupping were well known not only for the treatment of snakebite but for venomous bites and stings of other animals including the marine creatures.

B. Chemistry and Pharmacology

The marine toxins vary in their chemical and pharmacological properties far more than do the toxins of terrestrial animals. There is, however, some degree of consistency within a particular genus or species, although there are some notable exceptions. There also appear to be greater quantitative differences in some of the toxins of poisonous marine animals, such as in the puffer or ciguateric fishes, but here the differences can be accounted for. Under the circumstances, therefore, it is difficult to present a system of classification for the marine toxins in a manner that would allow reasonable consistency for both their chemical and toxicological properties, let alone be adaptable for clinical use. This is most evident with the sponge toxins (see Table 7).

Some marine toxins, for instance, are proteins of low molecular weight, while others are of high molecular weight. Some are polyethers, lipids, amines, alkaloids, guanidine bases, phenols, steroids, quaternary ammonium or halogenated compounds, or mucopolysaccharides, while the chemical nature of many is not yet known. In some marine organisms there may be several toxins present, and in some instances two organisms are necessary to produce a single toxin. Finally, it is well known that the toxin of an animal of one phylum may be identical to that found in an entirely different phylum. Such is the case of tetrodotoxin found in some newts, the pufferfishes, blue ringed octopus, Japanese ivory shell, a xanthid crab and others.

The pharmacological properties of the marine toxins vary as remarkably as do their chemical properties. Some marine venoms provoke rather simple effects, such as transient vasoconstriction or vasodilation, while others provoke more complex responses, such as parasympathetic dysfunction or multiple concomitant changes in blood vascular dynamics. The effects of the separate and combined activities of the fractions of these poisons, and of the metabolites formed by their interactions, are often further complicated by the response of the envenomated victim. The victim may produce and release several autopharmacologic substances, such as histamine, serotonin or other tissue components which may not only complicate the poisoning but which may in themselves produce more serious consequences than the toxin. There seems to be little doubt that in the evolution of marine venoms, as is the evolution of snake venoms (Russell, 1980b), synergistic and possibly antagonistic reactions may occur as the result of interactions between individual venom components.

In considering the biological properties of marine toxins it is well to keep in mind that there is no piece of experimental evidence to demonstrate that the chemical or pharmacological effects of a whole venom are equal to the sum of the properties of the individual fractions or functions. One of the unfortunate facts in the study of the chemistry and pharmacology of all the naturally occurring poisons is that the structure and design are most easily investigated by taking the poison apart. This has two shortcomings: first, it means that a destructive process must be substituted for a constructive, progressive and integrative one; and secondly, the essential quality of the whole toxin may be destroyed before one has made a suitable acquaintance with it. Sometimes the process of examination becomes so exacting that the end is lost sight of in the preoccupation with the means, so much so that in some cases the means becomes substituted for the end in the mind of the investigator.

Perhaps one of the most disturbing corruptions in the discipline of toxinology has been the blanket use of such terms as "neurotoxin," "hemotoxin," "cardiotoxin," "myotoxin" and "cytotoxin" when describing a whole venom. As previously indicated, venoms are complex substances having numerous pharmacological activities and synergistic reactions, and capable of inducing autopharmacological responses. Even with respect to individual fractions, it has also been shown that "neurotoxins" can, and do have cardiotoxic or hemotoxic activities, or both; "cardiotoxins" can have neurotoxic or hemotoxic activities, or both, and "hemotoxins" can have other activities. Until the individual fractions of venoms have been isolated and studied separately and in combination, we need to exercise extreme care in systematizing data which are based partly on biochemical studies, partly on biological assay methods, partly on clinical observations, and partly on hunches.

Of course, some venom or poison components may well have an effect at some specific site or membrane in the nervous system, but the basis for calling this fraction a "neurotoxin" is usually founded on its action on the particular preparation the investigator has selected, and rarely are the other possible pharmacological properties or tissue modulations studied. Today, our experimental techniques are sufficiently advanced to explore more specific pharmacological mechanisms at cell membranes, on cells or at specific tissue sites. We may find that the basic mechanisms involved are not site-specific but mechanism-specific, and act on several different kinds of tissues, whether the mechanism be ion or protein fluxes or other basic chemical or electrical phenomena. The shortcoming in transposing data from individual specific laboratory preparations to the intact animal, including humans, is that it can (and has) led to errors in deduction and, in the case of humans, to grave errors in clinical judgement. The scientist does not have the privilege of speculating too distantly from his data and must exercise extreme care when such data may need to be applied to clinical problems.

Although it is generally thought that a toxin which is only one component of a venom (a venom may have 100 other components, some of which are also toxic) is site-specific, that site or channel must be present in the prey or predator if the specific toxin is to be effective. Perhaps this explains why some animals are affected by a sodium channel blocker while others are not; the channel simply is not there. The literature sometimes fails to recognize such differences. This is particularly evident in comparisons between certain arthropods and mammals where channels can be quite different. It follows that if a toxin is to be useful to an animal, either in its offence or defence, its prey or predator must have a channel specific for that toxin. But bear in mind, most venoms are complex mixtures and may have many tissue sites, and that there may be components that have no observable tissue effects but which may act in synergism, or even antagonism.

This problem gives rise to selecting suitable bioassays. On examining the literature (see Russell *et al.*, 1984) one finds that the majority of pharmacological studies has been done on mammals, which may limit our understanding of the toxinological properties with respect to their role in the animal and the marine environment. However, since some of these studies are directed towards gaining knowledge for human therapeutics, they have that basic value. But how does one interpret such standardized tests as the LD_{50} in hemolytic terms, or vice versa? What is the significance of testing the intravenous LD_{50} of a toxin when the poison causes its effects following ingestion? What is the application of a neuromuscular block in the squid axon to patients who die in hypovolemic shock? Needless to say, all pharmacological data must be interpreted with care, and extrapolation must always be viewed with suspicion.

The modern period of marine toxinology must begin with the 1950s, and include such works as Phillips and Brady (1953), Buckley and Porges (1956), Kaiser and Michl (1958), Nigrelli (1960), Halstead and Russell (1964), Russell (1965a,b), and then by far the most important work, Halstead (1965–1970). Several of the recent texts, some of which have become more specialized, include those by Scheuer (1978), Hashimoto (1979), Pearn (1981), Russell (1984), Sutherland (1983), Williamson (1985) and Halstead (1988).

C. Clinical Problem

The clinical problem of stings and bites by venomous marine animals and the ingestion of poisonous fishes is probably far greater than generally supposed. Unfortunately, reliable statistical data on injuries and deaths caused by these animals are not available. Such records are not maintained in most countries and such fatalities are often included under "trauma," "accident," "poisonings," or some other classification, which often includes snakebite, plant poisoning, etc., as well as dog or other nonvenomous animal injuries.

Some statistics, however, are available. In 1971, one of us (FER) estimated that the number of injuries by venomous marine animals must exceed 40,000 a year. Fortunately, the number of deaths, excluding those due to sea snake venom poisoning, was small, probably less than 75 a year (Russell, 1971). However, since 1971 additional data on marine stingings indicate that previous estimates may have been far too low. For instance, Maretic (personal communication, 1980) reported that in 1978, 250,000 persons were stung by one species of coelenterate along the northern Adriatic coast of Yugoslavia. Letters from marine stations on the European and African coasts of the Mediterranean indicate that at least 1,000,000 persons are stung by cnidarians each year in that area. Updated records from the United States, Australia, South Africa, Hawaii, Tahiti and the Philippines would indicate that the number of stingings throughout the world must average 1,500,000 a year. As near as we are able to determine, the number of deaths from such stingings, excluding those caused by secondary infection, is still probably less than 75 a year.

2

PROTISTA

A. Introduction

In this chapter, as in the following, we have described those poisonous or venomous invertebrates that are known in the oceans around India. We have also noted many of the organisms that on a worldwide basis are toxic. This has been done because some of these creatures no doubt exist in the Arabian Sea or Indian Ocean, including the Bay of Bengal. Perhaps, these same organisms are not toxic in these waters, in particular with respect to the protistan, or their toxicity has not yet been demonstrated around India. In any event, the text has been designed to be a general reference work from which the interested student, researcher or physician can gain further insight into the toxic marine animals of the Indian Ocean.

The bioactivity of marine algae of Protista, particularly their antimicrobial effects, have been studied by a number of workers (Garber *et al.*, 1958; Burkholder and Sharma, 1969; Baslow, 1969; Hoppe *et al.*, 1979; Blunden *et al.*, 1981; Cocamese *et al.*, 1981, among others). Naqvi *et al.* (1981) and Rao *et al.* (1991) have adequately reviewed this subject as related to Indian waters. Further, it is not possible to review the tumorous or antitumorous, anti-inflammatory, antiviral, bacteriostatic or other properties of these toxins. Those interested in the overall range of biological properties exerted by marine animal substances should consult Thompson *et al.* (1991).

The term Protista is the kingdom taxon suggested by Ernst Heinrich Haeckel for the simplest of the not-quite-plant-not-quite-animal organisms. The word amounts to the claim that in this taxon are the "very first" living creatures (Jennings and Aker, 1970). Although the claim might justifiably be challenged, the term has survived and it is now used to denote those plant and animal unicellular organisms which have remained at the simplest grade of cellular construction. Among the protistan considered in this text are the various protozoans, algae, diatoms, bacteria, yeasts, fungi and organisms known as the cytoplankton and phytoplankton.

With respect to the marine forms and their toxins, the phytoplankton, and in particular those of the order Dinoflagellata, are the most important to toxinologists because of their relationship to paralytic shellfish poisoning (PSP). They also happen to be a valuable source of marine fauna food.

Dinoflagellates are eukaryotic organisms, principally free-living and motile unicells, although some are coccoid, filamentous or sac-like parasites. They are widely distributed throughout neritic waters and in the high seas from the polar oceans to the tropics. Some protistans such as *Gonyaulax (= Protogonyaulax = Alexandrium) catenella* and *Prorocentrum lima* are circumglobal. There are approximately 2000 species of dinoflagellates of which a good many are toxic to man and other animals (Table 1). Several species may be found in beach sand near the water's edge (Saunders and Dodge, 1984) or in tide pools or salt marshes. Dinoflagellates vary in size from 10 μ to 2 mm, with most being less than 100 μm in length. Their shape varies with the species and other factors. They are very fragile and propel themselves by means of two flagella which are lost at death or in fixing. At night, some dinoflagellates can be seen by their luminescence, lighting up the sea. As a group they grow slowly. *Gambierdiscus toxicus*, for instance, undergoes a division every three days. In its coral reef habitat this species is biflagellate, usually motionless and attached to microalgae, but it may swim when disturbed (Withers, 1984).

Blooms of Protista sometimes occur and result in the phenomenon commonly referred to as "red tide," "red water"

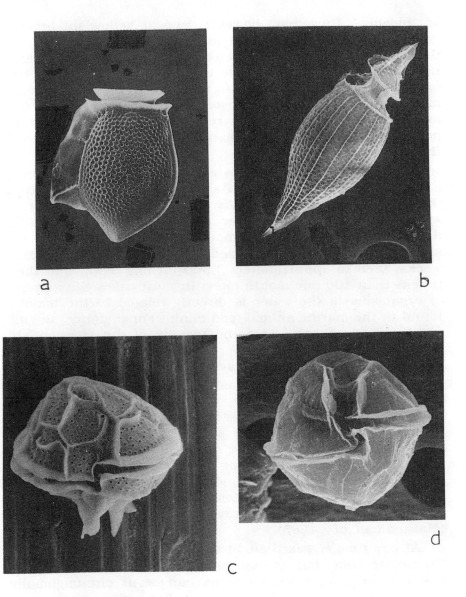

Figure 3. (a) *Dinophysis acuta* Ehrenberg. The girdle is virtually at the apex of the cell and the sulcus, with its extensive list, is along one of the narrow sides of the cell. 72 μm. 1, 55 μm. w. (b) *Oxytoxum sceptrum*. A short, stout species with a much reduced epitheca. 35 μm. 1, 17 μm. w. (c) *Pyrodinium bahamense*. A dinoflagellate often responsible for luminescence in the Indian Ocean and implicated in paralytic shellfish poisoning. 40 μm. 1, 60 μm. w. (d) *Alexandrium tamarensis*. Usually found very close to shore and often implicated in paralytic shellfish poisoning. 30 μm. 1, 35 μm. w. (Courtesy: J.D. Dodge.)

or "blooms." They may, however, appear yellowish, brownish, greenish, bluish or even milky in color, depending on the species and other factors. Their numbers may also vary. Blooms usually become visible when 20,000 or more of the organisms are present in 1.0 ml water. However, blooms may contain 50,000 or more organisms per ml, or as few as 15,000. The red color in red tides is probably due to peridinin, a xanthophyll. Radiant energy provides photosynthesis.

In an interesting study on a fringing reef of Tahiti over a period of almost 15 years, Bagnis *et al.* (1990) found that the total number of *Gambierdiscus toxicus* in a bloom varied anywhere from under 10 per gram algae to over 13,000 during a period of weeks, months and years. In one instance, their numbers increased from 50 to 13,000 in three weeks, then to approximately 3,000 for one week, and then to less than 100 one month later. In most cases the number of organisms in the water is directly related to the toxicity found in the marine animal food chain. For instance, Bagnis *et al.* (1990) observed that the toxicity of the surgeonfish, *Ctenochaetus striatus* (maito), was directly related to the blooms of the benthic dinoflagellate *Gambierdiscus toxicus*.

It has also become increasingly apparent that in those areas where PSP periodically recurs the responsible organisms might not be homogenous but "different sub-populations that may be varieties of one species or a group of closely related species" (Anderson *et al.*, 1990). The heterogeneity of *Alexandrium* populations, for example, indicates that there were 34 known strains found in a single area of over approximately 800 square kilometers of ocean, with toxicity being greatest at the northern reach of the organisms (Maranda *et al.*, 1985).

At one time it was thought that PSP was confined to the Temperate Zone, but as data have become more available it has become obvious that blooms can occur circumglobally in all zones. Since dinoflagellates are over 500 million years older than humans, blooms must have occurred prior to the advent of Man. Although this does not discount the problems of pollution caused by Man, it does raise questions on what role pollution contributes to toxicity at the present time. Red tides were known during biblical times and were certainly familiar to the North American Indians when pollution played little part in their settlements. Even during more recent times the least polluted waters off North America, namely the coasts of Maine and Alaska, are often the scene of massive blooms. On the other hand, increased nutrients emptied into an area of the sea may increase the number and growth of phytoplankton, but only further studies on this and other factors will determine their significance. The protistan implicated in poisonings to man or other animals are shown in Table 1.

The factors that contribute to the movements and numbers of Protista vary with the species and the number of different species in an area, their benthic resting cysts, with certain oceanographic conditions, high-tidal dissipations, upwelling of water masses, turbulence, surface illumination, temperature, salinity, runoff trace metals and organic nutrients, meteorological conditions, offshore shelf changes, thermocline development changes brought about by the seasons, degradation of stratification and other factors (Russell, 1965a,b; Halstead, 1965; Steidinger, 1975; Taylor and Seliger, 1979; Russell 1984; Ragelis 1984; Anderson *et al.*, 1985; Graneli, *et al.*, 1990).

Whenever excessive numbers of these unicellular organisms collect there may be mass mortality of fishes or other marine animals in the area (Carrilo, 1892; Kofold, 1911; Hornell, 1917; Nightingale, 1936; Galsoff, 1948; Connell and Cross, 1950; Fish and Cobb, 1954; Smith, 1954; Brongersma-Sanders, 1957; Grindley and Taylor, 1962; White, 1977; National Marine Fisheries Service, 1977; De Mendiola, 1979; Taylor and Seliger, 1979; Graneli, 1990). These mass mortalities may be due to a toxin produced by the protistans, or oxygen depletion in the water due either to the number of plankton present or to the release of decay products by these organisms and the dying animals, or to the mechanical effects of clogging of the gills. It should be noted that some of the recorded fish and other marine animal kills have been associated with non-toxic protistan.

The relationship between blooms of plankton and shellfish poisoning was perhaps first noted by Lamoureux (cited by Chevallier and Duchesne, 1851), who observed that during certain seasons of the year the sea appeared as a yellowish "foam" and that this foam was probably responsible for the poisonous properties of the shellfish. Most early workers,

Table 1. Protistan shown to contain or release a toxin that gives rise to paralytic, or neurotoxic, or diarrhetic shellfish poisoning through the food-chain; or that produces respiratory distress or dermatitis in humans; or that causes mass mortality of marine animals; or that have been implicated by laboratory experiments as being toxic. Ciguatera fish poisoning is not included.

Alexandrium acatenella (Whedon and Kofoid) Balech	PSP, toxic
A. catenella (Whedon and Kofoid) Balech	PSP, toxic
A. dimorpha (Biecheler) Taylor	PSP, toxic
A. fraterculus (Balech	PSP
A. fundyense (Balech) Taylor	PSP, fish kills, toxic
A. lusitanicum Balech	PSP, toxic
A. minutum (= *A. ibericum*) Halim	PSP, toxic
A. monilatum (Howell) Taylor	Fish kills, toxic
A. ostenfeldii (Paulsen) Balech and Tangen	PSP, toxic
A. polygramma Stein	Marine kills
A. tamarense (Lebour) Balech	PSP, fish kills, toxic
A. tamiyavanichi (Balech)	PSP, toxic
Amphidinium carterae Hulburt	Toxic
A. klebsii Kofoid & Swezy emend, D. Taylor	Toxic, fish kills?
A. rhychochepalum Anissimowa	Toxic
Anabaena flos-aquae (Lyngbye) Brebisson	Toxic, dermatitis, G.I.
Aphanizomenon flos-aqua (L.) Ralfs	Toxic, dermatitis, G.I. (?)
Caulerpa sp.	Fish kills, toxic
Chattonella antiqua (Hada) Ono	Fish kills
C. marina (Hada & Chihara)	Fish kills, toxic
Chlorella pyrenoidosa Chick	Toxic
Chysochromulina polylepis Manton and Park	Fish kills
C. leadbeater Esteb *et al.*	Fish kills
Cochlodinium catenatum O. Kamura	Fish kills, toxic
C. polykrikoides Margalef (= *C. heterolobatum*)	Toxic
Coolia monotis Meunier	Toxic
Dinophysis acuminata Claparède and Lachmann	G.I. distress, toxic
D. acuta Ehrenberg	G.I. distress, toxic
D. caudata Saville-Kent	G.I. distress, toxic
D. fortii Pavillard	G.I. distress, toxic
D. norvegica Claparède and Lachmann	G.I. distress, toxic
D. sacculus Stein	G.I. distress, toxic
D. tripos Gourret	G.I. distress, toxic
Fragilaria striatula Lyngbye	Toxic
Glenodinium foliceum Stein	Toxic
G. rubrum Whitelegge	Fish kills
Gloeotrichia echinulata (Smith) Richt	Toxic
Gymnodinium sp. (New Zealand)	NSP, G.I., respiratory irritant
G. breve Davis	NSP, G.I., NSP fish kills, toxic
G. catenatum Graham	PSP, toxic
G. flavum Kofoid and Swezy	Fish kills
G. galatheanum Braarud	Marine kills, toxic

Contd.

Table 1 Contd.

G. mikimotoi (= *G. nagasakiense*) Miyake and Kominani ex Oda	Marine kills, toxic
G. pulchellum Larsen	Fish kills, toxic
G. sanguineum Hirasaka	Fish kills, toxic
G. veneficum Ballantine	PSP, toxic
Gyrodinium aureolum Hulburt	Fish kills, toxic
Hemieurteptia antiqua Hada	Marine kills
Heterocapsa triquetra (Ehrenberg)	Fish kills
Heterosigma sp.	Fish kills
Linguldinium (= *Gonyalux polyedra)* Dodge	Toxic
Lyngbya majuscula Gomont	Dermatitis
Microcystis aeruginosa Kutzing	Toxic
M. toxica Stephens	Toxic
Noctiluca scintillans (Macartney)	Marine kills
Nodularia spumigena	Toxic
Nostoc spumigena (Mert.) Drouet	Toxic
Ochromonas danica Pringsheim	Toxic
O. malhamensis Pringsheim	Toxic
Oscillatoria nigroviridis Thwaits	Toxic
O. lenticularis Fukuyo	Toxic, PSP?
O. ovata Fukuyo	Toxic, PSP?
O. siamensis Schmidt	Toxic, PSP?
Peridinium polonicum Woloszynska	Fish kills, toxic
Phalacroma mitra Schütt	G.I. distress, toxic
Phalacroma rotundatum Claparède and Lachmann	Toxic
Polykrikos schwartzi Butschli	Marine kills
Prorocentrum sp.	Marine kills, G.I.
P. balticum (Lohmann)	PSP, fish kills
P. concavum Fukuyo	Toxic
P. cordatum (Ostenfeld)	Marine kills
P. hoffmannianum Faust	Toxic?
P. lima (Ehrenberg) Dodge	G.I. distress, toxic
P. mexicanum Tafall	Toxic
P. minimum (Pavillard) Schiller	Toxic
P. maculosum	Toxic
Pyrodinium bahamense var. *compressum* (Böhm) Steidinger, Tester and Taylor	
Prymnesium atelliferum Green	Fish kills
P. parvum Carter	Fish kills, toxic
Scenedesmus obliquus (Turp) Kutz	Toxic
Schizothrix calcicola (Agardh)	Toxic
Scrippsiella sp.	Toxic
Tubinaria ornata J. Aq	G.I.
Trichodesmium erythraeum Ehrenberg	Toxic

Modified from Steidinger (1983), Russell (1984), Taylor (1984, 1985), Shumway (1990), Steidinger (1993), and (personal correspondence, 1995).

however, attributed the poisoning to other causes such as copper salts, putrefactive processes, diseases of the shellfish, a "virus," other marine organisms, contaminated water, industrial wastes, bacterial pathogens, etc. PSP can occur where no bloom has been observed, although conjugates of protistans may not always be evident and, of course, blooms of toxic dinoflagellates may be recorded without the finding of PSP in a host animal.

In 1888, Lindner suggested a food-chain relationship for shellfish poisoning and subsequently this hypothesis received more favorable consideration. It was not until 1958, however, that Randall provided the convincing evidence for the food-chain relationship. In 1937, Sommer and Meyer published the results of their intensive investigation on the problem of paralytic shellfish poisoning. They demonstrated a direct relationship between the number of *Gonyaulax catenella* in sea water and the degree of toxicity in the mussel *Mytilus californianus*. These workers also established methods for extracting and assaying the poison, and suggested an experimental and clinical approach to the problem that served as a guide for many subsequent investigations.

In the Arabian Sea, red tides have been known since the 15th century. There are 14 reported species of dinoflagellates in Indian waters (Table 2), of which several have been suspected of toxic blooms. According to Subrahamanyan (1985), red tides are of common occurrence along the Indian coasts. Aiyar (1936) was perhaps the first to record widespread fish and invertebrate mortality associated with blooms of *Noctiluca miliaris* along the east coast of India. He suggested that the mass mortality was due to oxygen depletion in the water. In western Indian waters, *N. miliaris* was also responsible for mass fish mortality (Bhimachar and George, 1950) which appeared to be associated with putrefaction of the dinoflagellate, giving rise to thick masses of "slime that caused mechanical obstruction to the movement of fish, while its foulness was toxic to fish" (Subrahamanyan, 1985). However, no deaths were observed in fishes during blooms of *N. miliaris* in Park Bay but there was marked reduction in the presence of "herring," which was attributed to the decimation of the fish's food, primarily copepods and diatoms. Massive mortalities of fishes were also observed in 1957 and 1959. The incidence in 1959 was estimated to involve almost 20 million tons of fishes over an area of 1000

Table 2. Bloom-forming dinoflagellates in Indian waters.

West Coast	East Coast
Ceratium fusus	*Ceratium furca*
Ceratium macroceros	*Diplopsalis lenticula*
Ceratium tripos	*Gymnodinium* sp.
Dinophysis caudata	*Noctiluca scintillans*
Glenodinium lenticula forma asymmetrica	*Prorocentrum micans*
Noctiluca miliaris	*Protoperidinium pellucidum*
Peridinium depressum	*Scrippsiella sweeneyae*

(From Subrahamanyan, 1985)

× 200 km. On at least five occasions in 1986 and 1987 large masses of dead fish were observed (Rabbani and Nisa, 1990).

An outbreak of PSP in 1983 was confirmed by mouse bioassay (Karunasagar *et al.*, 1984) and the toxin subsequently studied on liquid chromatography. Gonyautoxins 1, 2, 3, 4 and 8 and 11-epigonyautoxin were the major toxins isolated. Small amounts of saxitoxin, neo-saxitoxin, decarbamoylsaxitoxin, decarbamoylgonyautoxins-2 and 3, C_3 and C_4 were also found. The offending organism was thought to be *Alexandrium* sp. In 1964, Prakash and Sarma observed a mass of *Gonyaulax polygramma* off Cochin and subsequently tested acid extracts of the organism in mice. The mice showed no significant signs and no deaths were recorded. It was observed, however, that fish avoided the protistan mass area, although they returned there several days after the bloom had drifted away.

In India, between 1984 and 1986, Karunasagar *et al.* (1989) took monthly water samples from major shellfish harvesting areas along the cost of Karnataka State in order to study the potentially toxic dinoflagellates. Although *Protogonyaulax* sp. were observed in two samples, the species were non-toxic. During April 1985 and April 1986, when low levels of PSP were detected in shellfish, no *Protogonyaulax* or *Pyrodinium* were observed in the water. *Dinophysis* and *Prorocentrum*, however, were detected and extracts of hepatopancreas of the shellfish produced death in mice, which they said resembled the manifestations of diarrhetic shellfish poisoning (DSP).

In the Indo-Pacific area the dinoflagellate chiefly responsible for PSP appears to be *Pyrodinium bahamense* var *compressa* (McClean, 1984). The commercially important shellfish frequently sampled for PSP are the clams (*Meretrix* sp., *Kalelysia* sp., *Paphia* sp.), the oysters (*Crassostrea* sp.) and mussels of the genus *Perna*. A mass mortality of fishes in the Arabian Sea off Gwadar Bay (Pakistan) was reported in November of 1987. *Prorocentrum minimum* was implicated in the bloom which extended over an area of 7 sq km and lasted for three days. The organism density ranged from 8–45 million per liter of water, and approximately 95 percent of the bloom was found to be *P. minimum*. Between 15 and 300 dead fishes were collected from the beach each day over the three-day period. The number of dead fishes that may have been washed away was not known. Eight species of fishes were involved in the deaths of which 59 percent were *Congresox* sp. ("pike conder"). The bloom was thought to be associated with high water temperatures, high light intensity and moderate winds.

Early in the 1980s it was proposed that various PSP toxins were produced by bacteria or in some way interacted with a dinoflagellate to produce the toxin. Since then, several workers have supported this concept. *Pseudomonas* sp. and *Vibrio* sp. isolated from their food protistan transformed gonyautoxin −1 −2 and −3 to saxitoxin by reductively eliminating N-1 hydroxyl and C-11 hydroxysulfate groups. Other bacteria have also been found to reduce the gonyautoxins to saxitoxin, suggesting a spectrum of bacteria that could take part in the reaction (Kotaki *et al.*, 1985; Kodama *et al.*, 1988). The bacterium *Moraxella* sp. was isolated from *Protogonyaulax tamarensis* and was thought to be the responsible agent in the toxin. Subsequently, it was found that *Moraxella* was a responsible contributor to the production of gonyautoxin, and that its numbers increased when under conditions of starvation in culture preparation (Kodama *et al.*, 1990). Some serious questions, however, have been raised and the role of bacteria in PSP, at present, is not clear. As noted elsewhere, PSP has been noted when no protistan have been observed in an area, and even when large numbers are present there has been no resulting toxicity in mussels or other animals. It has also been shown that several different kinds of protistan toxins can be found in a single animal. For instance, STX, neosaxitoxin, GTX-1, and -2, and a group of dissimilar toxins have been found in certain crabs (Llewellyn and Endean, 1989).

PSP has been obtained by direct extraction of the toxin(s) from the shellfish, from dinoflagellates secured from natural blooms and from laboratory cultures. In most cases, PSP can be obtained from all three sources in like form. Burke *et al.* (1960), Schantz *et al.* (1966), and Proctor *et al.* (1975) have grown *Gonyaulax catenella* in axenic cultures in cell densities equal to those occurring during natural blooms, and Schantz (1960) showed that the chromatographic properties of the toxin from the cultured organisms appear identical to those found in natural blooms and in mussels.

It has also been demonstrated that the toxicity of various dinoflagellates in culture can vary within a species, depending on nutrients and other chemical influences (Shimizu, 1978b). For instance, in one isolate of *Alexandrium fundyense* toxin, composition varied systematically with growth rate. In the presence of severe nutrient limitations cells grew slowly and toxin concentration was dominated by one or two toxin epimer pairs. As nutrients were increased, cell growth improved and the toxin profiles become more heterogeneous (Anderson *et al.*, 1990).

It is well known that the various protistan are capable of causing deleterious effects in fishes and other marine animals, as well as in humans and domestic animals by one or several mechanisms. In mass numbers they may deplete the oxygen content of the sea water or form strands or accumulations of protistan bodies that occlude the gills or breathing apparatus of other marine animals. They can poison a marine animal directly as in mussels, clams or oysters, or they can poison them indirectly through the food chain, as occurs in fishes. Some phytoplankton can produce multiple effects when the bloom is large enough. For instance, bloooms of *Gymnodinium breve* are known to kill fish, cause neurotoxic shellfish poisoning in oysters, and in its aerosol form they cause conjunctival and mucous membrane irritation in humans. Domestic animals have also been poisoned by PSP and CTX, and even whales have been found with saxitoxin-like poisons in their livers.

B. Paralytic Shellfish Poison (Poisoning), PSP

Paralytic shellfish poison or poisoning, known in various forms as saxitoxin (STX), neosaxitoxin (neoSTX), *Gonyaulax* toxin or gonyautoxin (GTX), gymnadin, dinoflagellate poison, brevetoxin, okadaic acid toxin (OAX), maitotoxin (MTX), or mussel or clam poison is a toxin or group of toxins found in some molluscs, arthropods, echinoderms and certain other marine animals which have ingested toxic protistan and have become "poisonous" through the food chain. When these animals are ingested by humans, PSP ensues. The problem has been discussed by Lindner (1888), Randall (1958), Russell (1965 a,b), Halstead (1965), Schantz (1971), Southcott (1975), Shimizu (1978b), and Hashimoto (1979). Since 1980 there have been at least 15 fine works on PSP, including those by Shimizu (1984), Anderson *et al.* (1985), Ragelis (1984), Graneli *et al.* (1990), Hokama (1991) and Shimizu *et al.* (1991). The interested reader is referred to these compendia for a more definitive review. In this section the authors will present a summary of what is now known about PSP and references as to where more detailed data can be found. We do this with apologies for errors of omission and commission.

1. Chemistry and Pharmacology

It is difficult to separate the chemistry of protistan toxins from its pharmacology, since chemists often use pharmacologic methods of bioassays for separating and identifying their fractions, and pharmacologists often employ chemical methods to obtain the toxin for their pharmacologic studies. The chemist provides us with what a substance *is*, while the pharmacologist or toxinologist tells us what it *does*. In this section, some attempt has been made to separate the two disciplines but as the reader will find, the two sciences are herein intermixed.

The impetus for the studies on the chemistry of PSP was a mass poisoning in humans at Wilhelmshaven, Germany, in 1885. Following this, Salkowski (1885) prepared alcoholic extracts from mussel tissues, concentrated them by evaporation and then reconstituted them in water. When the reconstituted product was forced into alcohol, a viscous precipitate formed. The filtrate from this precipitation contained the poison. Brieger (1888) isolated a substance he called "mytilotoxin," which produced effects in animals similar to those provoked by eating toxic mussels. Richet (1907) isolated a substance that caused changes in animals not unlike those described for one of Brieger's toxic fractions.

Since these signs were similar to those he had previously noted following poisoning with a toxin ("congestine") from a sea anemone, he called the new poison "mytilocongestine."

In 1922, Ackermann identified a number of bases in extracts from mussel tissues, including adenine, arginine, betain, neosin, methylpyridylammonium hydroxide and crangonine, none of which, however, had the properties of PSP. Partial purification of the toxin from the mussel *Mytilus californianus* was obtained by Müller (1935), who used permutit as an absorbant, eluted with saturated potassium chloride, and separated the poison by extraction of the residue from evaporation with methanol. Meyer *et al.* (1928), Prinzmetal *et al.* (1932), and Sommer and Meyer (1937) carried out more definitive investigations. Prinzmetal *et al.*, (1932) demonstrated that the poison from the mussel *Mytilus californianus* was slowly absorbed from the gastrointestinal tract and rapidly excreted by the kidneys. It was said to depress respiration, the cardioinhibitory and vasomotor centers, and conduction in the myocardium.

Sommer and Meyer (1937) were perhaps the first investigators to provide a complete report and approach to the problem of PSP, including the relationship between dinoflagellates and shellfish toxicity, extraction methods and assays, and an overall review of the medical problem at the time. This study still provides a protocol for many investigators who have limited experimental resources and experience. They found that 3,000 *Gonyaulax* weighed 100 µg (wet weight) and that this number yielded 15 µg of the dry extract, which in turn gave 1 µg of pure poison or one mouse unit. A mouse unit, or average lethal dose was defined as the amount of toxin that would kill a 20 g mouse in 15 minutes (Prinzmetal *et al.*, 1932; Sommer and Meyer, 1937). Thus, the amount of toxin contained in a single *Gonyaulax* was taken as 1/3000 of a mouse unit. Subsequently, various testing methods and assays were studied by Medcof *et al.* (1947), Meyer (1953) and McFarren *et al.* (1956) and the Association of Official Analytical Chemists (1975). At present, and with respect to *Protogonyaulax (= Gonyaulax) catenella* it is felt that the ingestion of 200-500 µg of the toxin is likely to cause some symptoms, while 500–2000 µg might cause moderate to severe symptoms, and more than 2000 µg could produce serious or even fatal poisoning. Variables in toxic properties have been noted by Bond and Medcof

(1957), Evans (1975), Schantz *et al.* (1975), and Dale and Yentsch (1978). Subsequently, Sommer *et al.* (1948) obtained the toxin following passage of extracts through charcoal, removing the lipid impurities with ether, passing the residue through sodium permutit, and then treating with ethanol. On evaporation, they obtained a product with a lethality of 6–12 μg per mouse unit. High yields of the toxin from California mussels and Alaska butter clams, using chromatography on weakly basic Amberlite XE-64 prior to chromatography on acid-washed alumina, were obtained by Schantz and his group in 1957. This clam is still the best source of STX.

In 1953, Fingerman *et al.* observed that saxitoxin had a marked effect on the frog peripheral nerve and skeletal muscle. They attributed the "curare-like" action to some mechanism that prevented the muscle from responding to acetylcholine. Bolton *et al.* (1959) obtained somewhat similar results. They demonstrated a progressive diminution in the amplitude of the end plate potential of the frog nerve-muscle preparation exposed to the poison. Murtha (1960) also noted that the toxin depressed mammalian phrenic nerve potentials, suppressed the indirectly-elicited contractions of the diaphragm and often reduced the directly-stimulated contractions. It was concluded that the effect of the poison was greater on reflex transmission than on the nerve. McFarren *et al.* (1956) presented oral lethal doses for eight animal species (Table 3).

Table 3. Oral LD_{50} per kilogram in mouse units for crude acid extracts of toxic clams, *Saxidomus giganteus* (McFarren *et al.*, 1956, 1960).

Animal	LD_{50}[a]
Mice	2100
Rats	1060
Monkeys	2000–4000
Cats	1400
Rabbits	1000
Dogs	1000 (approximately)
Guinea-pigs	640
Pigeons	500

[a] Mouse units (Sommer or International)

Ballantine and Abbott (1957) obtained a toxin from *Gymnodinium veneficum* that was soluble in water and dilute alcohol, insoluble in ether and chloroform, nondialyzable, and could be decomposed by hot alkali. Starr (1958) showed that *G. breve* toxin was similar to that from *G. veneficum*. Mold *et al.* (1957) found that distribution of the toxin in a solvent system of *n*-butanol, ethanol, 0.1 M aqueous potassium carbonate and α-ethyl caproic acid in a volume ratio of 146:49:200:5, with the aqueous layer adjusted to pH 8, resulted in a separation of the poison into two components, one of which was slightly more toxic than the other. At that time it was suggested that the poison existed in two tautomeric forms because upon standing in acid solution each of the components equilibrated to form the same mixture.

In 1955, a Canadian-United States Conference on Shellfish Toxicology adopted a bioassay based on the use of the purified toxin isolated by Schantz *et al.* Studies utilizing the methods outlined by the Conference indicated that the intraperitoneal minimal lethal dose of the toxin for the mouse was approximately 9.0 µg/kg body weight. The intravenous minimal lethal dose for the rabbit was 3.0–4.0 µg/kg body weight, while the minimal lethal oral dose for man was thought to be between 1.0 and 4.0 mg. Wiberg and Stephenson (1960) demonstrated that the LD_{50} of the then purified toxin in mice was:

Oral route	263 (251–267) µg/kg
Intravenous route	3.4 (3.2–3.6) µg/kg
Intraperitoneal route	10.0 (9.7–10.5) µg/kg

Subsequently, Schantz (1963) showed that both clam and mussel toxins were basic in nature, forming salts with mineral acids. The toxin was said by some to have a marked specific acetylcholinesterase inhibitory effect, similar to that of the organophosphorous compounds. This point, however, was not fully accepted. Schantz (1960) indicated that the contraction of isolated muscle in the presence of ATP and magnesium ions was not inhibited by the poison, nor did it alter the rate of oxygen consumption in the respiring diaphragm of the mouse.

At the Symposium on the Biochemistry and Pharmacology of Compounds Derived from Marine Organisms in 1960,

Murtha presented evidence that the toxin had a direct effect on the heart and its conduction system. He noted that it produced changes that ranged from a slight decrease in heart rate and contractile force, with simple P-R interval prolongation or S-T segment changes, to severe bradycardia and bundle-branch block, or to complete cardiac failure. He also demonstrated that it provoked a prompt but reversible depression in the contractility of isolated cat papillary muscle. In those cases where spontaneous contractions were abolished, Murtha found that the muscle still responded to electrical stimulation, although contractile force was reduced by about 50 percent. In both intact and partially eviscerated mammals, Murtha also observed a precipitous fall in systemic arterial pressure following injection of the toxin, indicating that the mechanism proposed by Kellaway (1935 a,b) (changes in the splenic circulation) was not responsible for the cardiovascular crisis. In vagotomized dogs, cardiac contractile force decreased 50 percent within the first minute following injection of the poison. There was a concomitant precipitous fall in systemic arterial pressure. In cervical cord-sectioned, bilaterally vagotomized mammals the immediate precipitous fall in arterial blood pressure was not seen, although some decrease in pressure subsequently occurred. The poison did not produce vasodilation in the vessels of the mammalian leg or kidney, nor did it affect the rate of blood flow in the isolated rabbit ear.

These various early studies on the action of PSP on the cardiovascular system indicated that the toxin had a direct effect on the heart, an effect which was in part responsible for the cardiovascular crisis. While the poisons may produce changes in the peripheral vascular system, these changes are probably not of sufficient magnitude to precipitate deleterious alterations in the systemic arterial blood pressure. It also appeared that a part of the cardiovascular crisis was in some manner related to the direct action of the poison on the central nervous system, although the experiments to that date did not exclude the possibility that cerebral anoxia secondary to cardiac-centered vascular failure may be a factor. Murtha (1960) suggested that the central nervous system effect may be mediated through the spinal cord.

In comparing the pharmacological effects of TTX and STX, Cheymol (1965) stated that the former was more active than the latter on the nerve but the latter was slightly more active on the neuromuscular preparation, even though its effects were more easily reversible. At the First International Symposium on Animal Toxins in Atlantic City, Kao (1967) demonstrated that the toxin blocked action potentials in nerves and muscles by preventing, in a very specific manner, an increase in the ionic permeability which is normally associated with the inward flow of sodium. It appeared to do this without altering potassium or chloride conductances. At the same meeting, Evans (1967) showed that in cats, mussel poison blocked transmission between the peripheral nerves and the spinal roots. The large myelinated sensory fibers were blocked by intravenous doses of 4.5–13 µg/kg, while the large motor fibers were not blocked until the dose was increased by about 30–40 percent.

In a subsequent paper, Evans (1968) observed that when dilute solutions of saxitoxin were applied locally to thin peripheral nerve branches in cats, conduction was not blocked. Conduction, however, was blocked in dorsal and ventral spinal root fibers following the topical application of far smaller concentrations. He suggested that one of the layers in the connective tissue sheath of peripheral nerve was impermeable to STX, while the leptomeninges covering the spinal roots were either deficient in or lack this layer. Subsequent work has shown that STX blocks ion transport by binding within and physically occluding the ion conduction pore of the sodium channel, but as pointed out by Catterall (1985) "there is little direct experimental support for this at present." STX has no appreciable affect on voltage-sensitive K^+ channels or on resting membrane conductances. These mechanisms appear similar for all components of the PSP complex, differing only in potency (Kao and Walker, 1982).

In 1970, Martin and Chatterjee isolated two toxins from laboratory blooms of *Gymnodinium breve* by acidifying the sea water and shaking it with chloroform. The first toxin (I) was found in the emulsion formed at the interface. It was separated and subsequently named "gymnodin" (Doig and Martin, 1973). The second toxin (II) was isolated by passing the chloroform extract through a silica gel column, washing with methylene chloride, eluting the residual material with ethanol and treating the product with charcoal. Its molecular weight was estimated to be 650 and its empirical formula $C_{90}H_{162}O_{54}P$. A toxin similar to substance II was isolated by exposing etheral extracts of *G. breve* cultures to thin layer chromatography on silica gel. Its elemental composition was C, 63.2; H, 8.9; O, 26.4; P, 1.2 percent, and a molecular weight of 468 was obtained by osmometry (Trieff *et al.*, 1972).

In 1971, Wong *et al.* of the Rapoport group presented evidence for the perhydropurine skeleton of STX by degrading it to aminopurine derivatives. Their final structure was:

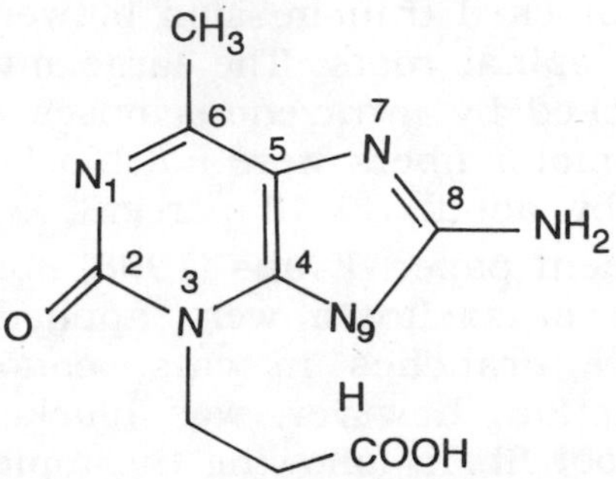

Further studies with a thoroughly dried sample gave $C_{10}H_{15}N_7O_3$.2HCl, rather than $C_{10}H_{17}N_7O_4$.2HCl Schantz *et al.* (1975) finally succeeded in establishing a crystalline di-p-bromobenzene sulfonate of STX and X-ray diffraction studies gave the structure which included the absolute configuration:

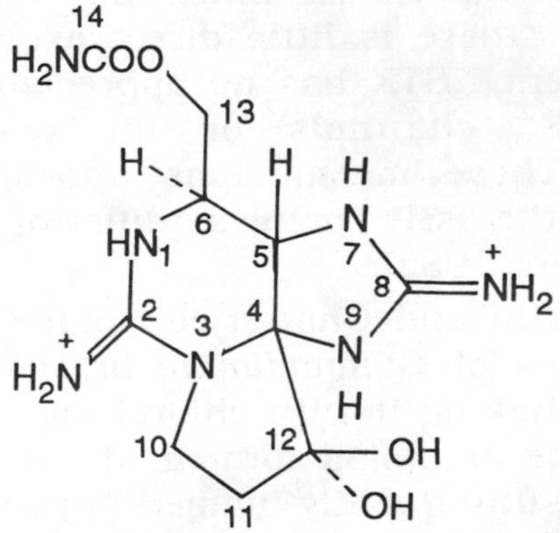

Working independently, Bordner *et al.* (1975), completed X-ray diffraction of a crystalline $C_{12}H_{21}H_7O_4$.H_2O obtained by treating STX dihydrochloride in an ethanolic solution. They confirmed the configuration as a stable hemiketal.

Shimizu *et al.* (1975) found that in the dinoflagellate *Gonyaulax tamarensis*, in addition to STX, there were several other toxins which differed from STX only in their weak binding ability on carboxylate resins. Further studies on organisms obtained from red tides along the New England coast resulted in the isolation of two new toxins, "gonyautoxin II" (GTX_2) and "gonyautoxin III" (GTX_3). Like STX, these new toxins lost their toxicity at high pH's. Both GTX_2 and GTX_3 were found to be highly hygroscopic. GTX_2 is a proposed 11-hydroxysaxitoxin, while the isomeric GTX_3 is considered a product of emolization. Subsequently, these compounds were found to be in the 11-O-sulfate forms (Boyer *et al.*, 1978; Shimizu *et al.*, 1981). Their structures are:

The Shimizu group developed a general isolation scheme in which the mixture of the toxins was separated by selective adsorption on Bio-Gel P-2 or Sephadex G-15, the toxic fraction eluted with dilute acetic acid, and the mixture applied on a column of weakly acid Bio-Rex 70. The acetic acid gradient elution yielded toxic fractions in the reverse order of the net positive charge of the molecule. The negatively charged toxins were not separable by this technique but could be separated by preparative thin-layer chromatography on Bio-Gel P-2. The interested reader is referred to the review by Shimizu (1991) for a more detailed accounting of the separation.

STX is a heterocyclic guanidine derivative. It has a molecular weight of 309 and a mouse intraperitoneal LD_{50} of 3.0 µg/kg. It is water soluble and very stable in acid solutions but unstable under alkaline conditions, particularly in the presence of oxygen. In general, the protistan toxins appear to be of lower molecular weight than the toxins of most other animals, many phytoplankton poisons varying from 300–3300 (Walker and Masuda, 1990). Some toxins, such as PSP and domoic acid, are water soluble, while others, such as CTX, brevetoxin and okadaic acid are fat soluble.

Another toxin, neosaxitoxin, a 1-*N*-hydroxysaxitoxin, has been isolated from *G. tamarensis* and exhibits chemical properties similar to STX, with one notable exception being that of its enhanced adsorption at 1770 cm^{-1}, which is attributed to a carbonyl functon (Shimizu, 1978). Its structure was confirmed by ^{15}N-NMR (Hori and Shimizu, 1983). It is not as stable as saxitoxin under acidic conditions. Its structure is:

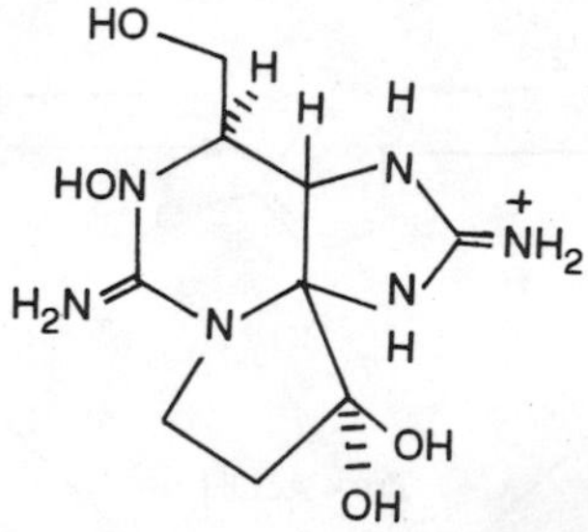

Domoic acid, which has been associated with amnesic shellfish poisoning, is a potent glutamate agonist that produces a continuous stimulation in neurons after binding to receptors in brain tissue (Collins, 1987). Its chemical structure is:

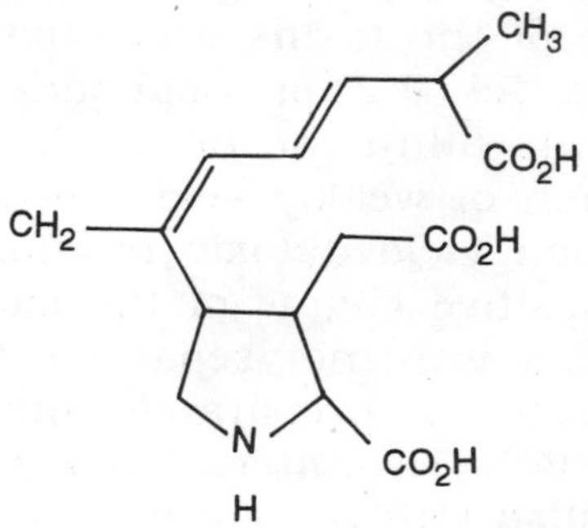

The toxin has been found in the blue mussel *Mytilus edulis* and suspected of being associated with blooms of *Nitzschia pungens* (Quilliam and Wright, 1989). In preliminary studies, Iverson *et al.* (1989) demonstrated that toxic mussel extracts were lethal at 2.4 mg/kg body weight in mice when given intraperitoneally. These levels were said to correspond to levels of 94 ppm in mammals. Oral doses between 35 and 70 mg domoic acid/kg body weight were needed to produce toxicity in mice and rats. Rats given 4.0 mg domoic acid/kg body weight developed bilateral edema in the arcuate nucleus of the hypothalamus and dark and pyknotic degenerated neurons in areas CA3 and CA4 of the hippocampus. Neuronal degeneration of the arcuate nucleus was observed in mice given 35 mg/kg domoic acid orally. Studies also showed that there was a lack of absorption from the GI tract: fecal excretion accounted for 98 ± 1–2 percent in animals that died. The authors concluded that since human intoxication occurred at an estimated 1–5 mg/kg body weight, susceptible individuals were more sensitive than rodents.

Brevetoxin (PbTx) and okadaic acid (OA) are considered separately from the other low molecular weight dinoflagellate toxins because of their unusual characteristics. PbTx is a group of poisons associated with red tides caused by the dinoflagellate *Ptychodiscus brevis*. It is responsble for the mass mortality of fishes (Steidinger *et al.*, 1973), causes upper respiratory and conjunctival distress (Music *et al.*, 1973), and through the foodchain is involved in human paralytic shellfish poisoning (Hemmert, 1975). There are at least 10 cyclic lipid-soluble polyether congeners in the complex, there being two types based on backbone structure (Baden *et al.*, 1989).

Brevetoxin A

Brevetoxin B

Brevetoxins are usually prepared from purified laboratory cultures using a combination of liquid extraction, thin
layer and flash chromatography, and high pressure chromatography. The major toxin, PbTx-2 is an aldehyde which
acts as a depolarizing agent, although at this time the exact
mechanism is unknown. The purified brevetoxin had been
shown to produce bronchoconstriction, as measured in
guinea pigs by pulmonary inflation resistance (Baden *et al.,*
1982a). When injected into mice, certain samples caused
characteristic muscarinic-like signs: hypersalivation,
rhinorrhea and increased urinary and gastrointestinal output (Baden and Mende, 1982). Atchison *et al.* (1986)
suggested that brevetoxin-B modifies a section of the sodium channel so that it opens at potentials more negative
than normal, and it inactivates to a lesser extent. This
would suggest a depolarization at the nerve membrane of
the neuromuscular junction and, in turn, explain the increased discharge of transmitter. Baden (1989) states that
the effect is excitatory through the enhancement of cellular
Na$^+$ influx.

In the frog and rat nerve-muscle preparation PbT-2 first
increases the rate of spontaneous transmitter release and
then produces a block (Gallagher and Shinnick-Gallagher,
1985; Atchison, *et al.*, 1986). In mammalian neuroblastoma
cells, PbTx-3 appears to increase sodium current at rest by
shifting the voltage dependency of channel activation, resulting in depolarization (Sheridan and Adler, 1989). In
conscious, tethered rats, PbTx-2 caused some degree of
heart block, premature ventricular contractions and

idioventricular rhythms, possibly through neurogenic control (Templeton *et al.*, 1989).

In rats, radiolabeled PbTx-3 was rapidly cleared from the blood. At one minute less than 10 percent remained in the blood and at 30 minutes, 69+ percent was found in muscle, 18 percent in the liver, and 8 percent in the intestinal tract. Fecal elimination accounted for 75 percent, urine 14 percent, and 9 percent remained in the carcass at six days. Poli *et al.* (1990) suggest that the liver is the principal organ of metabolism, and that biliary excretion is an important route of elimination.

An antiserum has been produced from the serum of goats for identifying PbTx at levels far lower than with conventional bioassays. The polyclonal antibodies were suggested to be useful in elucidating biosynthetic routes and as an immunoassay, although the antibodies did cross-react with extracts from *G. toxicus* (GTX) (Baden *et al.*, 1985). More recently, three new toxins have been isolated from *Gymnodinium breve* and known as hemibrevetoxin A, B and C. They have molecular weights almost half those of the brevetoxins (Prasad and Shimizu, 1989).

Okadaic acid (OA) is a polyether monocarboxylic acid found in the dinoflagellates *Prorocentrum lima* and *P. concavum*, and the sponges *Halichondria okadai* and *H. melanodocia* (Murakami *et al.*, 1982; Yasumoto *et al.*, 1989; Dickey *et al.*, 1990). It has a molecular weight of 786 and is relatively non-toxic, but it is responsible for diarrhetic shellfish poisoning that follows the consumption of shellfish contaminated by these protistan. Because of the almost circumglobal distribution of *P. lima*, poisoning of this type must be more common than generally believed. Its structure is:

Okadaic Acid (OA)

$R_1 = H$ $R_2 = H$

Several of its suggested mechanisms of action are thought to involve the release of arachidonic acid from cellular phospholipids and the stimulation of cyclooxygenase metabolites, but there may be other activities responsible in the mechanism of poisoning. Its possible role in ciguatera poisoning has also been noted (Dickey *et al.*, 1990).

In many studies on Protista, comparisons have been made on the basis of an intraperitoneal or even intravenous LD_{100}, the minimal lethal dose, or other unfortunately equally unreliable end points (Trevan, 1927; Russell, 1966a). On occasions, some of the basic principles of pharmacology have been overlooked, resulting in misleading toxicological premises (Russell, 1980a). Signs such as a mouse dragging its hind limbs (or even its tail) or jumping during an agonal period, often because of cerebral anoxia in the presence of a deficient blood supply to the brain (cerebral anemia), are frequently tagged as the effect of a "neurotoxin," and, as has so often occurred, another investigator finds the same substance to have an effect on red blood cell membranes or the heart and chooses to call it a "hemotoxin," or "cytotoxin" or even a "cardiotoxin." It is well to repeat that until we have studied the effects of a toxin on many tissues, as well as on intact preparations, care should be taken in labeling substances as neurotoxins, hemotoxins, and the like. This is not to imply however, that a toxin cannot be said to have neurotoxic, hemotoxic or other activities, but rather to express caution until we can differentiate between site and mechanism phenomena.

In mammals, the cause of death in PSP is probably related to respiratory dysfunction leading to ventilatory failure. When guinea pigs were subjected to PSP from *Gymnodinium* they developed a transient increase in airway resistance and a gradual decrease in mean inspiratory flow which led to elevated arterial blood gases and metabolic acidosis. These changes are thought to be related to a gradual failure of neuromuscular inspiratory drive (Franz *et al.*, 1989). It appears that STX is at least 10 times more toxic to mice by aerosol exposure than by systemic administration (Creasia and Neally, 1989).

In summary, these and other studies indicate that PSP blocks sodium channels, although the exact mechanism is still unknown. Shimizu (1991) has expressed it adequately when he states that the toxin molecule interacts with a receptor on the outside of the channel, possibly by hydrogen bonding and ion-pairing (see Shimizu, 1991, for references).

C. Other Protista Toxins

Some freshwater and marine blue-green algae are also known to be toxic to humans and to many animals (see Table 1). Among the animals that have been affected are man's domestic animals, fish, oysters, barnacles, coquinas, shrimps, crabs and zooplankton. For instance, Lightner (1978) has shown that blooms of the blue-green alga *Spirulina subsala* give rise to a high incidence of mortality in the blue shrimp *Penaeus stylirostris*. Death is caused by overwhelming necrosis of the epithelial lining of the midgut, dorsal caecum and hindgut gland, leading to hemocytic enteritis. It is believed that under certain conditions the alga produces a weak toxin that attacks the midgut.

A number of blue-green algae are of toxicological importance and because of their circumglobal distribution they are frequently contacted in marine waters. Since many of these are finely filamentous they often break up in shallow water, particularly when the sea is turbulent. In 1955, Habekost *et al.* reported that certain extracts of the blue-green alga, *Microcoleus lyngbyaceus*, sometimes termed "stinging seaweed," were toxic to mice. Arnold *et al.* (1959) described a contact dermatitis in bathers off Hawaii. They termed the lesions "seaweed dermatitis" and attributed the reaction to a marine alga. Grauer (1959) further described the dermatitis and it soon became known as "swimmers' itch," "surf itch" or "seaweed dermatitis." The disease was attributed to *Lyngbya majuscula* Gomont (now *Microcoleus lyngbyaceus*).

Banner (1959) found that the dried alga incorporated into pellets or a gelatin solution, killed mice when given orally. It should be noted that not all *Lyngbya* species are toxic and even the toxicity of *M. lyngbyaceus* can vary from locale to locale, and at different times of the year. Subsequently, the inflammatory agent of the alga was found to be debromoaplysiatoxin (Mynderse *et al.*, 1977), a substance which had previously been identified in the digestive gland of the sea hare *Stylocheilus longicauda*.

Debromoaplysiatoxin

Debromoaplysiatoxin was shown to produce a pustular folliculitis in humans and a severe cutaneous inflammatory reaction in the rabbit and hairless mice. The toxin is one of the most potent skin irritants known (Solomon and Stoughton, 1978). It has also been found in an algal mixture, which included *Oscillatoria nigroviridis* and *Schizothrix calcicola.*

In addition to the debromoaplysiatoxin and 19-bromoaplysiatoxin found in the mixture of *O. nigroviridis* and *S. calcicola*, several other toxic components, the most important of which was "oscillatoxin," have been identified (Mynderse and Moore, 1978). In addition to these substances, sterols, volatile constituents, fatty acids, malyngamides, pyrolic compounds, free amino acids and other components have been found in blue-green algae, but the specific pharmacological properties of these substances in the animals are unknown.

An additional vesicating substance has been identified in another shallow-water variety of *M. lyngbyaceus* found off Hawaii. This is an indole alkaloid, "lyngbyatoxin A," having the formula $C_{27}H_{39}N_3O_2$, and a mouse minimal lethal dose of 0.3 mg/kg body weight (Moore, 1977; Cardellina *et al.*, 1979).

It is known to produce escharotic stomatitis if eaten (Sims and VanRilland, 1981). Another blue-green alga, *Trichodesmium* sp., is known as "sea sawdust." It is a red, blue-green alga and is associated with massive fish kills. On

Lyngbyatoxin A

Guam, 13 people were recently poisoned following the eating of the red alga, *Gracilaria tsudae*; three died (Halstead and Haddock, 1992). As these authors point out, this incidence is of particular importance as Gracilariaceae are of considerable economic importance as food and for medical products. The nature of this toxin is unknown at present (Fig. 4).

The golden or yellow green algae *Chrysophyta* spp., are of particular importance because of the members *Prymnesium parvum, Ochromonas danica, O. malhamensis* and *Fragiolaria striatula*. *Prymnesium parvum* is a widely distributed phytoflagellate found in several oceans, in tide pools, estuaries, and in brackish waters. The first studies on its toxicity were carried out by Liebert and Deerns in 1920, following the mass death of fishes in the Workum-See. The organism was subsequently identified by Carter (1938). *Prymnesium parvum* became of further concern in 1957 when a bloom of the flagellate threatened the commercial fish-breeding industry in Israel. Subsequently, Droop (1954) in Scotland, and Reich and Kahn (1954) in Israel isolated and grew the organism in axenic cultures. In 1958, Yariv attempted to purify the toxin(s) from the organism and he and Hestrin (1961) suggested the name "prymnesin" to denote the extracellular toxin derived from water samples rich in *P. parvum*. The toxin was concentrated from culture filtrates and pond waters by adsorption on $Mg(OH)_2$. Further purification was obtained by treatment with acetone and methanol.

Figure 4. The red alga, *Gracilaria tsudae* (Courtesy: B.W. Halstead).

Shilo and Rosenberger (1960) extracted the toxin with methanol from dry cells concentrated by centrifugation. Paster (1968) purified a toxic prymnesin on concentrating the cells by centrifugation, removing the pigments with acetone and treating the crude prymnesin with methanol. The toxin was precipitated by ether, dissolved in water, and fractionated on Sephadex G-100. Following several additional steps, Paster obtained a toxin with a molecular weight of approximately 23,000 and elementary analysis of 42.5 percent C, 6.95 percent H, 50.55 percent O, and lipid constituents of about 30 percent, sugar constituents of about 70 percent, and an intraperitoneal LD_{50} for mice of 1.5 mg/kg body weight. It had a hemolysis index for rabbit erythrocytes of 25 mg/ml. Ulitzer and Shilo (1970), using a different separation technique, obtained a second toxin called "toxin B" which had 15 amino acids and a number of unidentified fatty acids. In contrast to prymnesin, toxin B resembled a proteolipid. It had six hemolytic factors.

Padilla and Martin (1973) also described the preparation of *P. parvum* toxin. Their toxin was isolated from the organism grown in 10 percent artificial seawater media maintained at 25°C under constant illumination. The sea water was enriched with liver fusion, glycerol and vitamins B_1 and B_{12}. After five to seven days growth of cells they were

collected by centrifugation, ground and the pigment removed with acetone. The acetone and insoluble residue were
extracted with methanol and placed on a Sephadex LH-20
column. The fractions were then collected and tested for
hemolytic activity.

In 1977, Kim and Padilla applied their *G. breve* separation technique for *P. parvum*. Using this procedure they
obtained six fractions, four of which were hemolytic and one
of these was toxic to fish but only in the presence of
spermine or at pH's lower than 9.0. The ability of the toxin
to hemolyse rabbit red cells is the assay most often used
for measuring the toxicity of prymnesin, as well as several
other protistan toxins. The assay was said to be 100 times
more sensitive than the ichthyotoxic test (Paster, 1973),
although it is not quite clear what the basic mechanism
might be in the ichthyotoxic test. The hemolytic activity was
said to be associated with membrane interaction leading to
lysis, causing loss of ions, amino acids and macromolecules
(Dafni and Gilberman, 1972).

The mechanism of the lytic action of *P. parvum* toxin has
also been studied by Imai and Inoue (1974). Using liposomes as a model membrane system they showed severe
damage to liposomes containing cholesterol but none to
those without cholesterol, which is rather different from the
findings of Ulitzer and Shilo (1970), who found that
Prymnesium toxin lysed spheroplasts or protoplasts of bacteria where the membranes lack cholesterol.

The crude *Prymnesium* toxin evoked a slow contraction
of the guinea-pig ileum, which was thought to be caused
by the release of acetylcholine, and the block of the contractions induced by smooth muscle stimulants (Bergmann
et al., 1964). It was said to have no effect on the rat uterus.
When studied on the frog heart the toxin caused a shortened action potential followed by a block in diastole with
marked depolarization. It was thought that the effects on
the frog heart and guinea-pig ileum were caused by the
same factor. In the frog sartorious nerve-muscle preparation
there was a decrease in the indirectly elicited contractions
on addition of the toxin to the bath. The directly elicited
contractions were little affected, and the nerve still propagated action potentials. End plate potentials disappeared
with time. It was suggested that the toxin acted as a non-

depolarizing blocking agent on the postsynaptic membrane of the endplate. In the deep extensor abdominalis medialis muscle (DEAM) preparation of the crayfish, a non-cholinergic junction, the block was also found to be at the neuromuscular junction (Parnas and Abbott, 1965).

In addition to the effects described above, the toxin alters a variety of mammalian and non-mammalian cells in cultures, bacteria, Ehrlich ascites cells and HeLa cells. It seems quite possible that the basic mechanism(s) involved in these tissue and cell changes is the same as that involved in the hemolysis; that the cytolytic, hemolytic, ichthyotoxic, neurotoxic, cardiotoxic, etc. changes may be due to a common mechanism.

The marine benthic green algae, Chlorophyta, are the organisms usually responsible for green tides or blooms. Several species are of importance to man as they provide part of his diet, particularly in the Philippines and other parts of the Orient. A usually edible benthic genus from the Philippines, *Caulerpa* spp., is known to be toxic during the rainy months, and injury to the plant thallus causes extrusion of toxin. The two toxins from *C. racemosa* and certain other species are known as caulerpicin and caulerpin (Aguilar-Santos and Doty, 1968; Maiti *et al.*, 1978).

$$CH_3(CH_2)_{13}-CH-CH_2OH$$
$$HN-CO-(CH_2)n-CH_3$$
$$n = 23, 24, 25$$

Caulerpicin

MeO$_2$C

CO$_2$Me

Caulerpin

These toxins are transferred through the foodchain to sea snails, soft corals, or by eating the dried alga. In humans the toxins produce paraesthesias about the mouth, tongue and terminal parts of the extremities, often as a feeling of coldness. Vertigo, ataxia and respiratory distress may also occur. The clinical manifestations are self-limiting and usually disappear within 12 hours.

The green alga *Cheatomorpha minima* is toxic to fishes and has hemolytic activity (Fusetani *et al.*, 1976). Another green alga, *Ulva pertusa*, also has several hemolytic fractions (Fusetani and Hashimoto, 1976), two being water soluble and one fat soluble. The fat-soluble hemolysin is palmitic acid, a C_{16} saturated fatty acid with a hemolytic activity of 0.24 saponin units/mg. One of the water-soluble hemolysins is thought to be a galactolipid, $C_{31}H_{58}O_{14}$, while the other is believed to be a sulpholipid, $C_{25}H_{47}O_{11}SK$ (Hashimoto, 1979). "Mozuku" poisoning is caused by several brown marine algae of the families Chordariaecae and Nemacystaceae. The poison may also be found in some red algae. Two fat-soluble fractions, "A and B", prepared from *Sphaerotrichia divaricata* and *Cladosiphon okanuranus* have been isolated as shown:

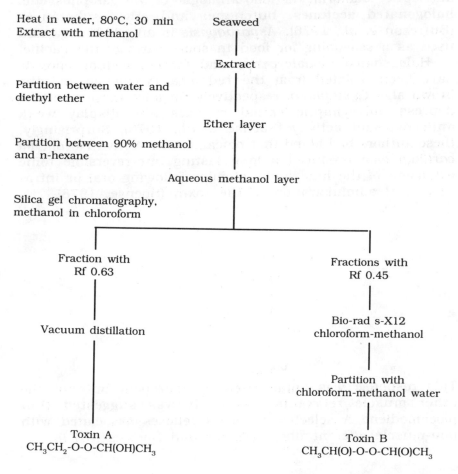

Heat in water, 80°C, 30 min   Seaweed
Extract with methanol

Extract

Partition between water and
diethyl ether

Ether layer

Partition between 90% methanol
and n-hexane

Aqueous methanol layer

Silica gel chromatography,
methanol in chloroform

Fraction with   Fractions with
Rf 0.63   Rf 0.45

Vacuum distillation   Bio-rad s-X12
chloroform-methanol

Partition with
chloroform-methanol water

Toxin A   Toxin B
$CH_3CH_2\text{-O-O-}CH(OH)CH_3$ $CH_3CH(O)\text{-O-O-}CH(O)CH_3$

The LD$_{50}$ of these toxins was not determined but toxin A had a mouse intraperitoneal lethal dose of 250 mg/kg body weight. The toxins are produced only on heating in water. Fusetani and Hashimoto (1981) suggest that the peroxides are responsible for mozuku poisoning, since the poisoning occurs after eating a hot water-infused preparation.

The red algae (Rhodophyta) and certain sponges contain numerous halogenated metabolites, particularly bromo compounds. The red alga *Asparagopsis toxiformis* contains many halo-compounds having 1–4 carbons. The major algal component is bromoform but the organism also contains the antiseptic, iodoform, the iodo analogue of war gas phosgene, halogenated acetones, butenones and other substances (Burreson *et al.*, 1976). *Asparagopsis* is an edible alga and used as a seasoning for food in some parts of the Pacific.

Halogenated monoterpenes and farnesylacetone epoxide have been isolated from the red alga *Plocamium* and the brown alga *Cystaphora*, respectively. In mice these fractions depress polysynaptic spinal reflexes and display weak anticonvulsant activity (Spense *et al.*, 1979). Surprisingly, these authors had found that plocamodiene A from *Plocamium cartilagineum* produced a long lasting but reversible tonic extension of the hind limbs of mice following oral or intraperitoneal administration of the toxin (Spense, 1978).

Plocamodiene A

This response was antagonized by diazepam but not the other muscle relaxants tested. It was suggested that plocamodiene A selectively affects reflexes associated with non-muscle afferent fibers (Taylor and Spense, 1979).

1. Assays for Protista Toxins

Assays for PSP have been made difficult because of the many different chemical and pharmacological properties of the various toxins, of which there may be more than 20 and by the fact that the amount of the toxin may influence the reliability of the assay. There is also the problem that the sulfamates of STX have low toxicity but they can be transformed to carbamates, which are much more toxic. The latter are usually the major components of the STX mixture found in the dinoflagellate. As many workers have pointed out, it is extremely important to differentiate the source toxins.

Perhaps the most valid indicator for detecting and quantitating the level of shellfish toxin in marine waters is the common mussel, since it has an almost worldwide distribution, is easy to obtain under most climatic conditions, can be collected in great numbers, its tissues are easy to dissect for quantitative determinations and the length of its periods of toxicity are fairly well established. In any monitoring program, however, such additional factors as the oceanography of the area, the motile cells in the water, and the benthic resting cysts, among other variables, need to be considered.

In 1964, Johnson *et al.* proposed the use of an immunochemical technique, while Bates and Rapoport (1975) described an analysis based on oxidation of STX to a fluorescent derivative. A spectrophotometric analysis was proposed by Gershey *et al.* (1977), and a unique cockroach bioassay was described by Clemons *et al.* (1980 a,b). Flow cytometric analysis of cellular STX, dependent on mithramycin fluorescent staining, was suggested by Yentsch (1981).

In 1984, Baden *et al.* proposed a polyclonal antibody technique as an assay against two major PbTx fractions in laboratory cultures. The method appeared to be both reliable and sensitive. The assay was also developed for enzyme-linked immuno-absorbent testing. Further studies comparing the immunoassay with molecular pharmacology techniques employing brain synaptosomes to study voltage-sensitive sodium channels, did not prove feasible (Baden *et al.*, 1988).

Chemically, the assay proposed by Sullivan *et al.* (1985) appears to be one of the most widely used at the present time. It is an efficient analytical system, using binary gradient HPLC and employing an adequate pump, a manual or automated injector system, a post-column reaction system, fluorescence detector with a xenon source and monochroamator, and a data system. Detection of the toxin using this system is based on oxidation of the toxins to fluorescent products. Its sensitivity and reproducibility are superior to the mouse assay, although at this time it is not possible to separate all of the PSP toxins in a single run; but by adjusting column conditions it is possible to isolate the known components in several runs (see Anderson *et al.*, 1985).

A comparison of the various assay methods for detecting shellfish poisons has been made by Hurst *et al.* (1985). In addition, antibodies have been raised in mice and shown to inhibit the binding of the toxin to membranes. This observation may be of value in diagnosis, as well as the treatment of the clinical case of poisoning (Huot *et al.*, 1989).

A fluorometric method for determining Okadaic Acid (OA) was devised by Lee and colleagues (1987). The technique proved accurate and more sensitive than the mouse bioassay. It was also time saving. An RIA has also been developed for the polyether. An immunogen was created by conjugating OA to bovine albumin with corbodiimide. Rabbits were immunized and antibodies produced. Sensitivity was at 0.2p mole. Specificity was demonstrated by showing that the OA antibodies did not react with maitotoxin, aplysiatoxin, palytoxin or brevetoxin. An enzyme-linked immunoabsorbent assay for detection of OA in diarrhetic shellfish poison has been developed and is now available in kit form (Uda *et al.*, 1989). Hauano *et al.* (1985) have suggested a bioassay using suckling mice for determining OA toxicity. An excellent manual on the toxins found in bivalves and the general consideration of shellfish sanitation with emphasis on India has been published by Ray and Rao (1984).

D. Clinical Problem

1. *Epidemiology*

With respect to humans, Halstead (1965) notes that there were 222 recorded deaths from PSP between 1889 and 1962. Prakash *et al.* (1971) note 80 reported cases around the Bay of Fundy from 1889 to 1970, with three deaths; and 107 cases in the St. Lawrence area, with 21 deaths during the same period. In California, between 1927 and 1989 there were 508 cases of PSP and 32 deaths (Price and Kizer, 1991). Global figures are not available but there may be close to 3,000 cases of paralytic shellfish poisoning each year.

These different figure estimates of poisoning reflect, as previously noted, the variables in the kind of shellfish, the quantity of toxin each harbors and how it is measured, the method in which the shellfish have been prepared, the size of the patient and other factors. In general, it is agreed that in spite of fairly well defined clinical manifestations there is some overlapping of symptoms and signs in most cases of PSP, and that most clinicians look to the most significant, predominant or first symptoms and signs in establishing a specific diagnosis. There is also the question of the quantity of the toxin, which may alter the clinical picture where two or more toxins are involved.

With respect to India, at Vayalus in the Chingelpet District at Tamil Nadu on the east coast, 85 people were poisoned after eating the backwater clam, *Meretrix casta*, from the Buckingham Canal. Three children less than 15 years old died (Silas *et al.*, 1982). Another outbreak occurred at Arikadu, near Kasaragod, northern Kerala in 1983 (Karunasagar *et al.*, 1984). In this incident, 15 individuals became ill after eating *M. casta* from Kumble Estuary, and a boy less than 14-years-old died in what was described as "respiratory paralysis." Unreported poisonings, however, are apparently known but the offending organism was not identified (N. Deoras, personal correspondence, 1965).

In the Vayalus episode the poisoning was attributed to dinoflagellates, although the organism was not identified. The comment was made that when the toxins were injected into mice they produced toxic manifestations, but it is not

clear if the toxins were taken from a sample of the consumed shellfish or from shellfish harvested in the same
area. The toxic dinoflagellate found in the Buckingham Canal
was *Gonyaulax polygramma* (Gopinathan, 1975). Because of
the high water temperature in both areas, it has been
suggested that the toxic dinoflagellate may have been
Pyrodinium bahamense var. *compressa*, or some unknown
warm water organism (Ray, 1984). Bivalve molluscan shellfish, particularly clams and mussels, are a regular food
source along the west coast and to a lesser extent in some
other areas (Alagarswami and Narasimham, 1973; Nayar,
1980; Silas *et al.*, 1982 and Ray and Rao, 1984). The last
mentioned authors highly recommended monitoring for sani-
tary quality, particularly where sewage pollution is high and
in those areas where previous outbreaks of PSP have occurred.

It has only been during more recent years that what was
once generally known as paralytic shellfish poisoning has
been separated into several distinct types of poisoning. This
has been done on the basis of the clinical manifestations
which, in turn, reflect the differences in the various toxins
involved in the poisoning, their quantities and the kind of
shellfish consumed. In reviewing the literature it also became apparent that the clinical syndrome is infrequently
described in medical terms. Perhaps, important symptoms
and signs have been missed, and in particular their chro-
nological development. It is surprising to find that during
the past five decades the number of truly clinical contributions on PSP represents less than 6 percent of the published
papers on the organisms and their properties.

As has been noted elsewhere for many kinds of animal
poisonings (Russell, 1980b), one of the important limiting
factors in PSP is that no physician sees or treats a sufficient number of specific types of cases to describe thoroughly
the clinical entity. One of us has seen or consulted on a
dozen such cases but even this is insufficient when it
comes to understanding the multiple mechanisms involved.
There is no doubt that many cases of PSP have been
misdiagnosed during times past, often being admitted to a
medical facility as "food poisoning." In the 1950s, for instance, one of us treated six patients in California that went
unreported as PSP. The first three of these were sent to the

Los Angeles Country General Hospital as "*Salmonella* poisoning." No doubt this was not an uncommon experience before the entity of PSP became well known and reportable. Frequently, such poisonings were associated with a heavy meal in which shellfish had been consumed, often with a history of "too much to drink."

Perhaps, one of the most thorough epidemiologic studies on PSP is that recently published by Price and Kizer (1991) on the problem in California. Their study covers the period 1927–1989 and involves 510 PSP cases, including 32 deaths. The California PSP Prevention Program is the oldest such program in the United States and is currently comprised of:

1. a mandatory reporting requirement for PSP,
2. an annual quarantine on sport-harvested mussels,
3. a coastal shellfish monitoring program,
4. PSP testing requirements for commercial shellfish harvesters, and
5. a public health information program.

Although this program has been very successful in controlling the occurrence of PSP, it is hampered by the seemingly changing and unpredictable dinoflagellate blooms. The 36-page report deals with the types of shellfish involved, their seasonal periods of toxicity and frequency, the intensity of the poisoning, and the range and dynamics of toxic dinoflagellate blooms. Also noted are differences in toxin levels in the mussels, oysters and the shellfish exposed to the same toxic bloom, commercial and sport shellfishing, and levels of illness reflecting ingestion of the poison. The compendium is highly recommended to anyone interested in the public health problem of PSP.

The amount of PSP that might cause a fatality varies considerably with the shellfish involved, the amount of toxin in that shellfish, the size and general health of the patient and other factors. In 1955, Tennant *et al.* found that severe manifestations of PSP could be seen with the ingestion of as little as 124 µg of toxin, and that death ensued following a 456 µg ingestion. Bond and Medcof (1957) cite a severe poisoning in a two-year-old following the ingestion of clams containing only 96 µg of toxin. In the same incident, two adults had mild symptoms and signs after consuming an estimated 340–344 µg of toxin, and a severe poisoning occurred from the ingestion of about 850 µg. In 1971, Prakash *et al.* suggested that a mild case of poisoning could be caused by the ingestion of 1 µg of toxin, which might be the amount found in 1–5 poisonous mussels weighing approximately 150 g each. A moderate case might be caused by 2.0 µg, and a severe poisoning by 3.0 µg, but these amounts are dependent on the kind of basic toxin. Evans (1975) notes that the manifestations are generally not felt unless more than 160 µg are ingested, while over 160 µg is necessary for a severe poisoning. Schantz *et al.* (1975) proposed that severe poisoning, and even death, may occur following 500–1000 µg of the toxin, while Dale and Yentsch (1978) place moderate poisoning at 1000 µg of toxin, with a lethal dose at 10,000 µg.

In a case seen by one of us, 12 mussels, *Mytilus californianus*, caused sufficient poisoning to require positive pressure assistance, cardiac stimulation and parenteral fluids with high doses of calcium. Based on our experience, mostly with mussels along the Pacific coast of North America, approximately 400 µg of PSP will produce some toxic manifestations. Serious or severe symptoms and signs are seen following the ingestion of 1000 µg, while 3000 µg or so may produce death in an adult to whom no treatment has been given. More recently, Price and Kizer (1991) have suggested that in California 200–500 µg of toxin is likely to cause at least mild symptoms and signs, 500–2000 µg moderate to severe poisoning, and the consumption of more than 2000 µg is likely to produce serious to lethal consequences.

In 1990, Taylor presented a very simple and practical clinical division of the various forms of phytoplankton poisoning:

	Known since
Paralytic Shellfish Poisoning (PSP)	1793
Diarrhetic Shellfish Poisoning (DSP)	1978
Neurotoxic Shellfish Poisoning (NSP)	1970s
Amnesic Shellfish Poisoning (ASP)	1987
Ciguatera Fish Poisoning (CFP)	16th century

This classification should go a long way in helping to clarify the different types of poisoning by protistan.

In 1991, several dozen people who had eaten the clams they had dug up as the Olympic Peninsula in the state of

Washington, USA developed nausea, vomited, complained of abdominal pain, subsequent short-term memory loss and headache. The illness was traced to the presence of domoic acid in the clams. Further south prior to this episode a number of deaths in pelicans and cormorants had been attributed to toxic dinoflagellates.

2. Symptoms and Signs

The first manifestations of PSP often begin 15–160 minutes following eating the offending shellfish and they are fairly consistent, though their appearance and severity are dependent upon the amount of toxins ingested. There is a paraesthesia of the lips and tongue which may sometimes involve the entire face and even the fingertips. The patient often describes this feeling as "tingling" or "numbness," even like "mild electric shocks." Vertigo or dizziness often follow and there may be a headache, nausea and even vomiting. "Stomach cramps" are frequently reported and subsequent diarrhea is not unusual. A general feeling of weakness develops which sometimes leads to partial paralysis, muscle incoordination and ataxia. The patient may complain of a "floating in air" sensation and mild respiratory distress. Diplopia, dysarthria and diminished or absent reflexes may be present, and in the more severe cases paralysis, usually ascending in type, develops. This may lead to respiratory failure and death. Most of the deaths on record occur between two and 14 hours following ingestion of the poisonous molluscs.

In both series of cases in India noted previously, the poisoning was paralytic in nature. In the former, the symptoms and signs included tingling of the lips, tongue and finger tips; numbness in the limbs; back pain; floating in air; blurred vision and gastrointestinal disturbances. In the latter series similar manifestations were reported, numbness of the lips occurring within 30 minutes, followed by vomiting and paresthesia and then paresis leading to paralysis of the limbs. In this latter series all victims recovered within 48 hours, except the one boy.

Diarrhetic shellfish poisoning usually occurs following the eating of scallops or mussels. The initiating and usually most troubling manifestations are gastrointestinal in nature and sometimes confused with "food poisoning" or an allergy reaction. The most common dinoflagellates associated with this form of poisoning are *Dinophysis* spp., and probably some species of *Prorocentrum*. The worldwide distribution of the former make it an important clinical syndrome. It differs from PSP in that there are either no neurological findings or they are very minor and not disabling. As the name implies, the most troublesome sign is diarrhea, which may begin within two hours of consuming the toxic molluscs, but generally four-to-six hours. In most cases there is abdominal pain, often severe, and nausea and vomiting may occur. The poisoning is self limiting but fluid balance must be watched. In one patient seen by one of us, dehydration and weakness were observed.

Neurotoxic shellfish poisoning occurs following the eating of mussels or clams in which the causative dinoflagellate is *Gymnodinium breve*. It is typified by tingling of the lips, sometimes the face and occasionally the tips of the fingers. Gastrointestinal manifestations are usually minor, although diarrhea may occur. Reported cases in North America have occurred during March or November, and involved shellfish that had been steamed. Incubation time varied from 30–130 minutes. There have been no known deaths from NSP.

A clinical syndrome, amnesic shellfish poisoning, was reported in 1990 by Perl *et al.*, following the consumption of toxic mussels harvested on Prince Edward Island, Canada. The poisoning was characterized by vomiting, abdominal cramps, diarrhea, headache, disorientation, confusion, loss of balance and loss of memory. The poisoning differed from

Table 4. Manifestations of mussel poisoning in 99 patients following consumption of mussels*.

Manifestation	No. of Yes Responses	Total Responses	%
Nausea	75	98	77
Vomiting†	74	97	76
Abdominal cramps†	48	95	51
Diarrhea†	41	97	42
Headache	40	93	43
Memory loss†	24	96	25

* The results were obtained from the standardized questionnaires completed for 99 of the 107 patients. Total responses do not add to 99 because not all questions were answered for each patient.
† Criterion for inclusion as a case (From Perl *et al.*).

PSP in that the principal neurological manifestations were short-term memory loss, confusion and disorientation. Of the 107 persons intoxicated by the mussels that met the diagnostic criteria, three elderly patients died and 12 of 19 hospitalized cases were placed in intensive care. All of the seriously ill patients and those that died were elderly. Table 4 shows a breakdown of symptoms and signs.

The issue of the importance of domoic acid as the principal causative factor in the neuropathology of the poisoning has been questioned by Snodgrass (1990) and some points of his argument have yet to be answered. Some investigators have suggested that perhaps not all of the manifestations (fever, for instance) could be attributed to the domoic acid, noting the possible presence of an endotoxemia. They discussed their findings in septic shock patients complicated by multiple-organ failure but Teitelbaum *et al.* (1990) questions this etiology for the poisoning. Other reports on amnesic shellfish poisoning will be found in the reports of Todd, 1990; Sutherland *et al.*, 1990; Zatorre, 1989; Glavin *et al.*, 1990; Olney *et al.*, 1990; Johnson *et al.*, 1990; Grimmett *et al.*, 1990; Bruni *et al.*, 1991, and others.

3. Treatment

In spite of the massive amount of literature on PSP, the scarcity of clinical data, particularly therapeutics, is remarkable. At this time one can only recommend that all patients be treated symptomatically: gastric lavage, if indicated; intravenous fluids for dehydration and hypotension; respiratory assistance to prevent respiratory failure and acidosis; oxygen, if necessary; the appropriate drug for nausea and vomiting but not during the first hour following the poisoning when vomiting might be encouraged, and lots of reassurance. The use of anticholinesterase drugs, β blockers and other specific agents used for neurotoxic poisoning do not appear to have been tried, but some of the basic neurophysiology with PSP indicates they might be of value.

3

PORIFERA

A. Introduction

Sponges are among the simplest of animals. They are highly organized acellular free-living organisms composed of loosely integrated cells covered by an epithelium and, with few exceptions, supported internally by a skeleton of silica and/or calcite, or spongin. They are devoid of digestive, excretory, respiratory, endocrine and circulatory systems and lack neurons. There are between 5000 and 5500 described species and an unknown number of undescribed species. Sponges are found in almost every sea from mid-tide levels to the deepest parts of the oceans. Most species form microscopic, inorganic spicules that are either calcareous or siliceous, and collagenous fibrils and fibers, the latter called spongin. Some horny species elaborate quantities of spongin. The skeletons of some of these have been used since ancient times because of their capacity to absorb fluid. Members of the family Spongillidae are found in freshwater to 3350 m above sea level. Reaggregation involves eggs and sperm and less frequently asexual reproduction by budding fragmentation, gemmules and the eventual production of asexual larvae, coalescing into successively larger groups until a small functional sponge is formed (Burton, 1961, Bakus, 1971). This technique has been used in attempting to cultivate sponge toxins, some biologically active components of sponges, and to study antigenic makeup.

Table 5 shows the Porifera that have been implicated as toxic. Our knowledge of the toxic sponges in the sea around India is relatively restricted. The most widely studied area for sponges has been the Gulf of Mannar (Burton, 1930; and Rao, 1941). In more recent years, Thomas (1983) has made a more detailed survey of the sponge fauna of the "Indian region." He noted 481 species of which 427 belong to Demospongiae.

Some sponges release a toxic substance into their environment. De Laubenfels (1932) observed that when *Tedania toxicalis* de Laubenfels was put in a bucket with fishes, crabs, molluscs and worms, in an hour, or perhaps less, the animals would die. Although this phenomenon has usually been considered as a purely defensive reaction initiated when the sponge became endangered, Green (1977) suggested that the toxic material may be released as a continuous product into the surrounding water, and thus serve as a warning or deterrent to an approaching predator. This seems reasonable. The idea of continuous elaboration of the toxin has elicited specific studies including those by Janice Thompson at the Scripps Institute of Oceanography who, working with *Leiosella* sp. has found that this sponge does secrete the toxin as a continuous process (unpublished results). It may be that quantitative differences in the release of toxin are involved, perhaps initiated by some physical or chemical stimuli.

In the boring sponge, *Siphonodictyon coralliphagum*, of the Caribbean, the animal secretes a mucus that dribbles down its sides and kills corals (Schulte and Bakus, 1992). While there seems to be little doubt that sponge toxins have a defensive function, Bakus (personal communication, 1982) suggests that, in addition to the fish or other predator repelling functions, the toxin may prevent the settling of larvae or spores on the animal's surface, and it may also prevent overgrowth on one species by another.

Much of the work on the ecology of toxic sponges has been carried out by Bakus and his colleagues at the University of Southern California (see Bakus, 1969). In his various works he has suggested that as the diversity and biomass of fishes increases, as in tropical waters as compared to temperate or polar waters, there is increased competition for food which, in turn, gives rise to

Table 5. Some marine sponges which have been reported to be toxic to humans (A) or whose toxic properties have been studied (B).

Name	Distribution
(A)	
Agelas Dispar	
Agelas oroides	
Biemma saucia	Indo-Pacific
Dysidea etheria De Laubenfels	West Indies, Florida
Ephydatia sp.	Australia
Haliclona viridis (D. & M.)	West Indies, North Carolina, Florida, ?West Central Pacific
Hemectyon ferox (D. & M.)	West Indies
Ircmia felix (D. & M.)	North Carolina to Venezuela, ?Mediterranean
Microctona prolifera (Ellis & Solander)	Atlantic coast of North America, Willapa Bay, Washington; San Francisco Bay, California
Neofibularia mordens	Circumglobal
Neofibularia nolitangere (D. & M.)	West Indies, Florida
Pseudosuberites pseudos Dickinson	Gulf California and Mexico
Spheciospongia vesparium (Lamarck)	West Indies, Florida
Spinosella vaginalis (Lamarck)	West Indies
Spirastrella inconstans (Dendy)	Indo-Pacific
Suberites domunculus (Olivi)	Mediterranean, Black Sea, Northwest Africa
Tedania digitata	Cosmopilitan
Tedania ignis (D. & M.)	
Tedania nigrescens (Schmidt)	Cosmopolitan
Tedania toxicalis De Laubenfels	California
(B)	
Adocia sp.	
Agelas dispar D. & M.	West Indies, Florida
Callyspongia sp.	
Callyspongia fallax D. & M.	West Indies, Florida, ?South Australia
Calyx nicaeensis (Risso)	Mediterranean
Chondrilla nucula Schmidt	West Indies, Florida, Adriatic Sea, Ligurian Sea
Dysidea etheria De Laubenfels	West Indies, Florida
Fasciospongia cavernosa	
Geodia sp.	
Geodia cydonium (Jameson)	Mediterranean, European seas, West Coast Africa
Geodia gibberosa Lamarck	North Carolina, West Indies, Guiana
Geodia mesotriaena Lendenfeld	Alaska to Gulf California

(contd.)

(*Table 5. contd.*)

Name	Distribution
Halichondria okadai (Kadota)	Japan, Korea
Halichondria panicea (Pallas)	Cosmopolitan
Haliclona sp.	
Haliclona doria De Laubenfels	West Indies
Haliclona erine De laubenfels	West Indies, Brazil
Haliclona rubens (Pallas)	Florida, West Indies, Gulf Mexico
Haliclona viridis (D. & M.)	Florida, West Indies, North Carolina, Gulf Mexico, ?West Central Pacific
Hemectyon sp.	
Hymemacidon ?amphilecta De Laubenfels	Florida, West Indies, Gulf Mexico
Hymemacidon perlevis (Montagu)	Mediterranean, East Atlantic, Indian Ocean. ?Australia, ?Japan
Ianthella sp.	
Iotrochota birotulata (Higgin)	Florida, West Indies, Indo-Pacific
Ircinia campana (Lamarck)	Gulf Mexico, Florida, West Indies, Brazil
Ircinia felix (D. & M.)	N. Carolina, Florida, Gulf Mexico, West Indies, Caribbean, Venezuela, ?Mediterranean
Ircinia strobilina (Lamarck)	Florida, West Indies
Keratose sponge	
Latrunculia magnifica	
Lissodendoryx aff. *kyma* De Laubenfels	California to Washington
Microctona parthena De Laubenfels	California
Microctona spinosa Wilson	Florida, West Indies
Mycale sp.	
Mycale lingua (Bowerbank)	Alaska to Washington and Newfoundland to Maine
Niphates erecta (D. & M.)	N. Carolina, Florida, Caribbean
Petrosia seriata	
Phyllidia varicosa	
Pseudoceratina crassa (Hyatt)	Florida, West Indies
Reneira saria	
Ridleia sp.	
Sigmadocia sp.	
Smenospongia sp.	
Spongia officinalis L.	Mediterranean
Tethya actinia De Laubenfels	West Indies, ?Marshall Islands
Tethya aurantia (Pallas)	Cosmopolitan
Tethya crypta	West Indies
Thalyseurypon sp.	
Xestospongia muta	
Xestospongia subtriangularis (Duch.)	Florida, West Indies

From Russell, 1965; Halstead, 1965; Hashimoto, 1979 and Russell, 1984.

exploitation of food sources. This leads to more specialized feeding habits. Such activity would appear to stimulate some type of natural selection in sponges, such as chemical defense through retardation or as in some cases, prevention of predation or grazing by fishes; diverse cryptofauna, animals hiding from fish predators among coral branches and the like: or changes in structure, etc. These among other mechanisms might be involved in the natural selection process by which the sponge survives. The natural selection of chemical defenses was subsequently demonstrated by Bakus (1981), who found that 73 percent of all exposed common reef invertebrates on the north Great Barrier Reef were toxic to fish. Sixty percent of the exposed sponge species were toxic, while only 33 percent of the cryptic species were toxic.

It is well known that a few fishes belonging to highly specialized teleost families feed on toxic porifera and are not affected (Randall and Hartman, 1968). These species also consume large quantities of non-toxic sponges. Perhaps, the process of detoxification in these fishes is more highly developed. It has been suggested that toxic sponge species, being eaten in only small numbers or amounts, are thus protected from rapid total depletion or may be part of the cryptofauna. In his work on fishes, Green (1977) demonstrated that the toxicity of sponges increased with decreasing latitudes and that the most commonly exposed sponges were the most toxic to fishes. The same is true of certain other marine invertebrates (Bakus and Green, 1974). Some nudibranchs feed exclusively on both toxic and non-toxic porifera without deleterious effects (Graham, 1955) as do, perhaps, some annelids and crustaceans. The feeding relationships between these animals and sponges is not well understood.

B. Chemistry and Pharmacology

To humans, many sponges have an offensive odor and taste but what part these qualities play in defense in the animals' biota is not known. Table 6 shows some of the marine sponges of medical or chemical importance. The reader should consult Halstead (1965), Jakowska and Nigrelli (1960), Stempien *et al.* (1970), Green (1977), Bakus and Thun (1979)

Table 6. Some marine sponges of the Indian Ocean which ·have been reported to have antimicrobial, antifouling or hemolytic activity.

Adocia pigmentifera	*Microciona rhopalophora*
Agelas ceylonica	*Mycale mytilorum*
Agelas sp.	*Mycale grandis*
Aulospongus lubulatus	*Phyllospongia dendyi*
Axinella domnani	*Phyllospongia foliacens*
Callyspongia fibrosa	*Sigmadocia fibulata*
Callyspongia diffusa	*Sigmadocia pumila*
Dysidea herbacea	*Spirastrella inconstans*
Dysidea fragilis	*Spirastrella inconstans* var. *digitata*
Geodia perarmata	*Spirastrella cuspidifera*
Geodia lindgreni	*Spongia officinalis* var. *ceylonensis*
Haliclona expansa	*Tethya robusta*
Haliclona pigmentifera	*Tethya japonica*
Haliclona viridis	*Tedania anhelans*
Ircinia fusca	*Verongia lacunosa*
Ircinia racemosa	*Verongia crassa*
Laxosuberitus cruciatus	*Xestospongia* sp.
Microciona prolifera	

and Stonik and Elyakov (1988) for more detailed data on the toxicity of these species. It should be noted that some sponges of the same genus as those found to be toxic to fishes have also been found to be non-toxic, and some species have also been found to be both toxic and non-toxic. The family Haliclonidae appears to have the most consistently toxic species.

Of the 481 species of Porifera noted by Thomas (1983), for the "Indian region," 35 are reported to have some sort of toxic properties (Sarma *et al.*, 1991; Rao *et al.*, 1991; Gulavita *et al.*, 1991; Bheemasemkara and Rao *et al.*, 1991). In reviewing the literature on the marine toxins from sponges, including those found around India, it becomes difficult to separate the toxic properties of the sponges from their physiological or pharmacological properties as toxins. With respect to the bioactive compounds from marine organisms, the reader will find the text by Thompson *et al.* (1991) of value, particularly the contributions by Gulavita *et al.*, and Rao *et al.* relating to the Indian oceans. Table 6 shows a list of Porifera found in Indian waters that are said to have antimicrobial, antifouling or hemolytic activities.

The list of bioactive substances in sponges reads like an encyclopedia (Russell, 1971). Table 7 shows some of the names given components that have been isolated from Porifera that are said to be "biologically active," having antitumorous, anticarcinogenic, anti-inflammatory, antiviral, antibacterial, bacteriostatic, vasodilatory, vasoconstrictor, tumor promoting, immuno-suppressing, or immuno-enhancing properties. As one writer once said, "A sponge a day keeps the doctor away." In the list are also substances that are said to be hemotoxic, myotoxic, neurotoxic, cytotoxic, anticytotoxic and cardiotoxic, although the pharmacological assays implicating these substances might be questioned. No attempt has been made to categorize these or place them in any specific order, since there is considerable overlapping and some duplication, some terms are now obsolete, and the specific chemical nature of most is not sufficiently clearly established to classify them.

The interest in the chemical structure of sponges arises from several factors: (1) they are the source of novel chemical compounds; (2) their chemical constituents have a potential as tools in the study of biological mechanisms; (3) they are a rich source of primitive organic compounds and secondary metabolites; and (4) they have been shown to have potentially useful products for humans. In this book we will deal only with those sponges and tissue components that, in contact with man or other animals, produce deleterious results.

One of the difficulties in the study of toxins from sponges has been the lack of a standardized applicable and reliable assay. The original observation by de Laubenfels (1932) was that when fishes and other marine animals were exposed to sponge secretions they either died or displayed deleterious manifestations. More recently this observation has been refined by Bakus and his colleagues. Essentially, the shipboard preliminary assay method now used by this group involves grinding 5 g of the sponge in 10 ml of sea water, centrifuging, pouring the supernatant into a bowl with 300 ml of sea water, and then placing a 1.5–5 g sergeant major (*Abudefduf saxatilis*) in the water and observing the fish's behavior over a designated period of time. In preliminary testing in the laboratory a similar program is followed, except that 200 ml of tap water and 1.2–2.4 g goldfish (*Carassius auratus*) are used.

Table 7. Some substances isolated from sponges having sundry properties.

Acetylcholine	Diketosteroid
Acid polysaccharides	Dimethylhistamine
Acridine alkaloids	Diterpenes
Adenine	Eicosadieonic acid
Aerophobin	Eledonine
Aeroplysinin	Ethyl-6-bromo-3-indolcarboxylate
Agelasine	Glycocyamine
Ageline-A	Glycoproteins
Agmatine	Guanidine
Amino acids	Guanine
Aragupetrosine-A	Halicholactone
Araguspongine-A-B-C-D-E-F-G-H,J	3-hydroxyacetal-6 bromoindole
Avarol	p-hydroxybenzaldehyde
Batzelline A,B,C	Haliciamine-A
Betaine	Halitoxin
Biogenic amines	Halogenated compounds
Bishomosesterterpenes	Herbipolin
Brominated tyrosine	Hippospongine
Bromoindoles	Histamine
Bromopyrrole derivatives	Homarine
Calyculins	Hypotaurine
Carotenoids	Imidazole alkaloids
Calysterol	Isocyanoneopupukeanane
Cholestanol	Isoquinolinequinone
Cholesterol	Isosarain-1
Choline	Isothiocyanatopupukeanane
Clathridina	Inosite
Clinasterol	Inositol
Collagens	Isoprenoids
Corallistine	Jasplakinolide
Cycloartenol	Lanosterol
Debromagelaspongin	Latrunculins
Dendrillolides A,B	Lectin
Dercitin	Lestecterpenes
Descoderminian	Lipids
Desoxyhavannahinis	Luffolide
Desoxyribonucleic acid	Lysine
Dibromoagelaspongin	Macrolides
Didemnis	Maoalide
Dienoic acids	Mercaptohistidines
Dihydroxyaerothionin	Methyladeine

(*contd.*)

(*Table 7 contd.*)

Methylcholesta	Quinone
Methylisoguanosine	Renieramycins
Methylpurine	Rhizochalin
Mycalamide A,B	Saraine-A
Mycalamide-A	Savinholida A
Myosin ATPase	Secoxestovanin A
Neohalicholactone	Sesquiterpene isotheocyanates
Neospongosterol	Sesquiterpenoids
Nerthyosides	Sesterpene
Norditerpens	Spongian
Nucleosides	Spongothymidine
Okadic acid	Spongouridine
Onnamide	Squalene
Ovothiol	Sterols
Oxaquinolizidine	Stylostatin
Pentosides	Taurine
Petrocin	Taurobetaine
Phosphoarginine	Taurocycamine
Phosphocreatine	Tedanoid
Phospholipids	Terpenoids
Phosphonosphengoglycolipid	Terpine
Plakorin	Tetramine
Polyhydroxysterols	Theonellaminde-F
Poriferasterol	Thymine
Prenylated benzoquinones	Ubiquinone
Purines	Uracil
Putrescine	Xestovanin-A
Pyridine	Zooanemonine
Pyrroloimidazoles alkaloids	

(Updated from Russell, 1971)

After preliminary testing, a standard laboratory bioassay is used. Serial dilutions of the crude extract are prepared, started with a 0.1 g crude material/ml tap water and an alcohol extract is used for comparison. Water temperature in the 11-liter aquarium is 22–27°C and the LC_{50} determinations are done for the more highly toxic sponges, using an alcohol extract. Toxicity is determined on the basis of the fishes swallowing air, blowing bubbles, being bitten by normal fish, equilibrium loss, erratic swimming behavior, slow swimming movements, escape responses, thrashing behavior, extreme lethargy or stupor, failure to recover when put in fresh water, and death (Bakus and Thun, 1979). In their study of 54 species of Caribbean sponges, it was found that 31 species were toxic to fish. The most toxic species was *Hymeniacidon amphilecta* de Laubenfels, with species of *Haliclona* being less toxic and *Ireinia* sp., Microcronidae, *Neofibularia nolitangere*, *Xestospongia muta* also being less toxic.

The first toxicity test, however, was carried out by Richet in 1906. He precipitated a substance from extracts of the siliceous sponge *Suberites domunculus*, which when injected into laboratory animals produced vomiting, diarrhea and dyspnea, and caused hemorrhages in the gastric and intestinal mucosa, peritoneum and endocardium. The lethal dose in dogs was 10 mg/kg, and the toxic substance was found to be non-toxic when administered orally. The poison was called "suberitine." Arndt (1928) demonstrated that extracts from certain freshwater sponges produced diarrhea, dyspnea, prostration and death when injected into homoiothermic animals. These extracts had some hemolytic effect on sheep and pig erythrocytes, and blocked cardiac function in the isolated frog heart preparation. The extracts were heat stable and produced no deleterious effects when taken orally. In 1951, Bergmann and Feeney isolated two nucleosides, spongouridine and spongothymidine from the Jamaican sponge *Tethya crypta*.

Spongouridine   Spongothymidine

These nucleosides occur free in sponges and have served as models for synthesizing the chemical analogue D-arabinosylcytosine, a nucleoside having antiviral and certain other pharmacological activities that have been attributed to its property of inhibiting pathways in nucleic acid biosynthesis (Cohen, 1963). Free amino acids have been identified in abundance in sponges, including some relatively rare naturally occurring ones, such as β-amino isobutyric and pipecolic acids (Bergmann, 1962; Berquist and Hogg, 1969; Berquist and Hartman, 1969). The free amino acids have been of particular importance in establishing fingerprints for taxonomic problems within the Porifera.

In 1971, Das *et al.* found that extracts of *Suberites inconstans* produced a histamine-like effect on the guinea-pig intestine and attributed this to histamine, which they found in the sponge. On paper chromatography they detected five other amines, three having phenolic groups. Dried specimens of *Fasciospongia cavernosa* yielded crystals of *N*-acyl-2-methylene-β-alanine methyl esters. In mice, the subcutaneous lethal dose of the crystals was approximately 120 mg/kg body weight (Kashman *et al.*, 1973), obviously not very toxic by toxicological standards. Agelasine from *Agelas dispar* has some activities of a saponin. A unique sesquiterpene, 9-isocyanopupukeanane, has been isolated from the nudibranch *Phyllidia·varicosa* and has been found to be present in the sponge *Hymeniacidon* sp., on which the nudibranch feeds (Burreson *et al.*, 1975).

Sesquiterpene

The sequiterpenes, and some isonitriles and isothiocyanate sequiterpenes are known to be toxic to various animals on injection. The reader is referred to Hashimoto's (1979) text for a more thorough review of the properties of these substances.

Berquist (1978) states that the modern period of reinvestigation of sponge sterols dates from 1972. She notes among the novel sterol types the astysterols, with new side chain alkylation patterns; the calysterols, having a cyclopropene ring in the side chain; and the norsteranes, with a modified tetracyclic ring structure. The sponges which have a moderate to high proportion of lipids in the form of sterols contain relatively small amounts of terpenes, and the reverse condition frequently applies. Although the sterols of sponges may not be particularly toxic, if at all, some investigators have suggested that they may play a part in the metabolism of the toxic component(s).

Berquist (1978) also notes that while sponges contain numerous biologically active substances, four groups are of particular interest: the terpenoids, heavily halogenated dibromotyrosine-derived compounds, bromopyrrole derivatives and the phenylated benzoquinones. The terpenoids are, for most part, linear furans, such as the antibiotic, "furospongin-1" and the sesterterpene, "furospongin-3" from *Spongia officinalis*.

Furospongin-1

Furospongin-3

The terpenoids are highly astringent to the taste and may cause the sponge to be unpalatable to predators.

The dibromotyrosine-derived compounds are small molecules containing bromine. *Verongia* sp. are of particular interest because of their potent antibiotic activity (Cimino *et al.*, 1975). Their toxicological properties have not been studied. The bromopyrrole compounds have been found in at least five species of three genera, *Agelas*, *Axinella* and *Phakellia* (Cimino *et al.*, 1975). "Oroidin," obtained from *Agelas oroides* and certain other species is a more complex bromopyrrole metabolite. Its toxicological properties are not

known. The phenylated benzoquinones of sponges are a
novel group whose pharmacological properties are of interest because related compounds in other marine invertebrates
function to confuse the olfactory sense of predators (Kittredge
et al., 1974).

In 1973, Wang *et al.* demonstrated that a preparation of
an extract of *Haliclona rubens* exerted a depolarizing action
on the end-plate membrane of the frog skeletal muscle and
that a lesser depolarization occurred in the membrane elsewhere than at the end plate. This activity differed from that
caused by batrachotoxin and grayanotoxin (a toxin from the
plant Ereaceae). The crystalline toxin from *Halichondria
okadai*, okadaic acid, has been isolated and characterized.
It is a cytotoxic monocarboxylic acid with the molecular
formula of $C_{44}H_{68}O_{13}$. Its intraperitoneal LD_{50} in mice is 0.19
mg/kg body weight. Spectroscopic studies indicate it is an
ionophoric polyether and diffraction crystallography of the
o-bromobenzyl ester indicates a heptacyclic structure
(Tachibana, 1980). Halitoxin, a toxic fraction from the genus
Haliclona (Green, 1977) was shown to be a biopolymer with
1,3-dialkylated pyridine as a repeated subunit. The toxin
was further studied by Schmitz *et al.* (1978) having separated it from three toxic fractions with molecular weights of
500–1000, 1000–25000 and >25000. One compound was
hemolytic, cytotoxic and lethal to mice. Halitoxin depolar-
ized end plate and muscle membrane potentials.

Cariello *et al.* (1980) isolated and characterized Richet's
suberitine. Living specimens were squeezed, the extract
centrifuged in the cold at 27000 × g for one hour, the
sediment resuspended in sea water and centrifuged again,
the two supernatants exposed to cold ethanol, and the
precipitate removed by centrifugation. The sediment was
then extracted with sodium acetate, the suspension centrifuged, and the precipitate re-extracted with the same buffer.
The procedure was repeated until the supernatant displayed
no lethal activity; concentration was by ultrafiltration. The
product was next fractionated on Sephadex G-150 and the
active fraction, which caused paralysis when injected into
the crab, was chromatographed on Sephadex G-75 and
further purified on Sephadex C-50. Homogeneity was demonstrated on SDS polyacrylamide gel electrophoresis, and
the molecular weight, approximately 28,000 was estimated

by ultracentrifugation. The toxin had a marked hemolytic effect on human erythrocytes and some ATPase activity. Studies on the giant axon of the abdominal nerve of the crayfish showed that in a concentration 4.4 mg/ml there was depolarization followed by an irreversible block in the indirectly stimulated action potential. The authors speculated that the irreversible block may explain the flaccid paralysis seen in crabs following injection of suberitine into the arthropod's hemolymph.

Quinn *et al.* (1980) found that a purine nucleoside extracted from *Tedania digitata* caused marked muscle relaxation, reduced blood pressure and blocked spinal polysynaptic responses in the mouse. *Latrunculins A* and *B* have been isolated from a fluid exuded by the branching red sponge *Latrunculia magnifica* of coral reefs in the Red Sea (Groweiss *et al.*, 1983). These toxins caused hemorrhage in mice and fish by altering the organization of microfilaments without affecting the microtubular structure. Schmitz *et al.* (1984) also isolated a potent cytotoxin, tedanolide, from *T. ignis*. The great brown sponge, *Petrosia seriata*, was shown to have antifeeding activity for fishes (Braekman *et al.*, 1982), while a component of the Mediterranean sponge, *Reneira sarai*, has shown moderate activity in mice (Cimino *et al.*, 1986). The interested reader should consult Braekman and Dalose, 1986; Endo *et al.*, 1986 and Stonik and Elyakov (1988) for further data on the properties of sponge extracts.

C. Sponge Poisoning

With respect to humans, poisoning probably occurs through deposit of the toxin(s) in the very superficial abrasions produced by the fine, sharp spicules of the sponge. It has been known since the days of Pliny that traumatic injury to the human skin can be produced by the spicules, and it is believed that in many cases of poisoning this occurs prior to the deposit of the poison on the skin. Certainly, an abraded skin is more likely to absorb a toxin than an uninjured one. In 1965, one of us reported a case of poisoning involving a 27-year-old skin diver, who having abraded the skin of his hands while collecting coral, decided to assist in the packing of fresh *Tedania nigrescens* aboard the boat. After handling these sponges for approximately 30 minutes, he complained of an intense burning sensation over the hands, pruritus and malaise. When seen an hour later, the pain was described as intense and the palms of the hands were intensely red.

In 1942, Cleland reported the Australian sponge, possibly *Ephydatia* sp., caused dermatitis on handling. Subsequently, Yaffee and Stargardter (1963) reported an interesting case and series of experiments relating to the stings of *Tedania ignis*. The patient noted a burning, itching erythema of the right palm shortly after handling the sponge. The redness and itching lasted six to eight hours, but 10 days later there was a recurrence of the burning and itching, and during the subsequent 48 hours elevated pruritic, erythematous-ringed lesions appeared on the volara of his arms; his fingers became red and swollen as did his face and soles. In another of their cases (Yaffee, 1963), a skin diver picked up *T. ignis* and within 10 minutes complained of localized burning, itching, and transient erythema which lasted for about 15 minutes. Two weeks later the patient presented with mottled erythematous eruptions over his hand, which was slightly swollen, and enlarged epitrochlear and auxiliary lymph nodes.

In their studies on the Australian sponges *Neofibularia mordens* and *Lissodendoryx* sp., Southcott and Coulter (1971) noted the handling of the former sponge resulted in subsequent stinging pain, erythema and swelling over the contact areas. Some cases were initiated by itching, then the development of a localized rash and subsequent desquamation. They noted that specimens frozen up to four years remained toxic, and even specimens immediately dried often retained their ability to produce some clinical manifestations, although dry sponges were non-toxic. They stated that "wetting is necessary to activate the toxic agent." Toxicity was reduced on boiling or immersion of the sponge in a soap solution. *Lissodendoryx* sp. was also a contact irritant but less so than *Neofibularia mordens*. Das *et al.* (1971) reported contact with *Suberites inconstans* resulted in localized pruritis and minor swelling.

Sims and Irei (1979) reported a stinging by *Tedania ignis* in Hawaiian waters. The patient developed redness over the hands, burning and tenderness with subsequent purpuric mottling. Vesicles developed and peeling occurred a week later. This paper provides an excellent review of the problem of sponge poisoning.

In 1985, while diving with the Human Underwater Biology group in shallow water off Bonaire, a 14-year-old male plunged his diving knife into what he subsequently described as a "large brown sponge" and was immediately engulfed by a "white milky fluid" extruding from the sponge. Swimming around the sponge he noticed a "sort of stinging sensation" over the arm and hand holding the knife. He immediately surfaced and it was observed that he was hyperemic over his entire body. When first seen by one of us, his skin was definitely hyperemic. He was anxious, restless and hyperventilating, and complained of burning "like poison oak" over his face, arms and trunk. His body was washed with cold salt water and salt water soap, dried and "Itch Balm Plus" (hydrocortisone 0.5 percent, diphenhydramine 2 percent and tetracaine, 1 percent) applied. This afforded relief. Further examination revealed a BP of 110/60, pulse 92 and respirations 36. Other clinical findings were normal. He was kept warm and in a supine position for four hours by which time the "burning" had almost ceased. He urinated three times during this period. He complained of weakness or fatigue and was put to bed. The next morning all symptoms and signs had dissipated.

In 1991, Hooper *et al.* reported a severe contact dermatitis following the handling or exposure to the desmacellid *Biemma saucia*. They also noted an exudate discharged by the sponge and suggested that the exudate also contained spicules which might be able to penetrate the skin or even through thin neoprene wet suits and gloves. Analysis of the sponge revealed a chemical irritant, p-hydroxybenzaldehyde, which they suspected of being the toxic substance.

The typical clinical syndrome on contact with a toxic sponge is localized pain and itching at 10–20 minutes, followed by a burning or prickling sensation, the development of redness or a rash, tenderness, localized swelling on some occasions, and vesicle formation and desquamation in the more severe envenomations. These are the usual self-limiting complaints following sponge injuries. The regional lymph glands may be enlarged and tender. When the affected area subsequently comes in contact with any object, a prickling sensation may occur, even hours after the initial

A

Figure 5. *Tedania ignis* is probably the most dangerous sponge with respect to handling injuries.

(A) Live specimens (Courtesy: B.W. Halstead).

B

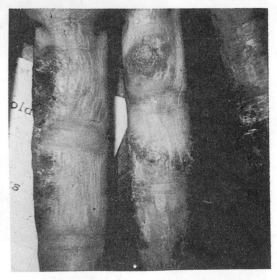

C

(B) Dead specimen (Courtesy: R. Cameron).
(C) Lesions at one week following handling of *Tedania ignis* (Courtesy: W. Ornis, Scripps Collection).

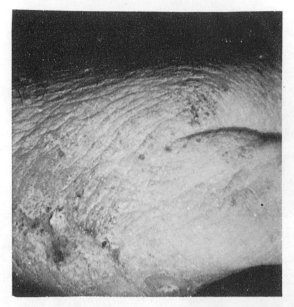

a

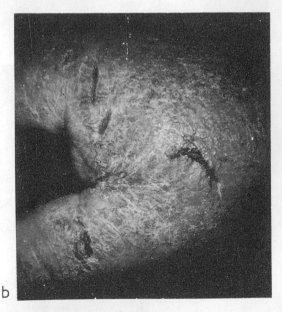

b

Figure 6. Localized effects of sponge injuries.

(a) Fire sponge lesions showing typical eczematous changes
    (Courtesy: A.A. Fisher).

(b) Sponge injury to hand two weeks following the handling of a toxic
    sponge (Courtesy: A.A. Fisher).

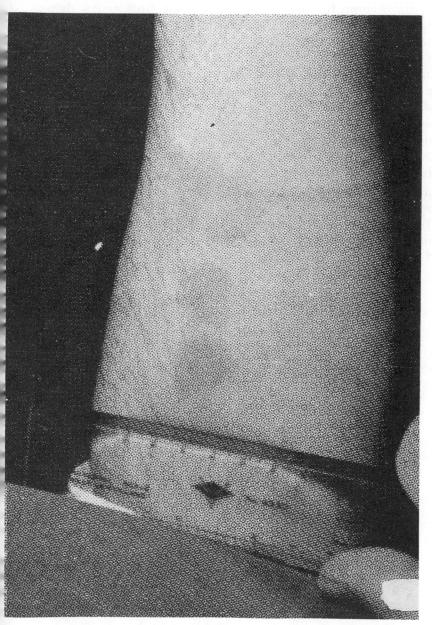

Figure 6c. Quantitated experimental lesions at three days from brief exposure to the stinging sponge. *Neofibularia mordens.* Left, internal surface of sponge; center, external surface; right, sponge exposed to formol-sea-water for 12 hours. (Courtesy: R.V. Southcott).

contact with the sponge. The symptoms and signs may last for a few hours to, as in the case of the dermatologic lesions, many months. If the affected area undergoes desquamation, it may remain pigmented for years. It can be suggested on the evidence provided by Hooper *et al.* (1991), and the case described by us, that contact with the offending sponge is not always necessary. The animal appears to be able to discharge an exudate into the surrounding sea water which contains a toxin (p-hydroxybenzaldehyde?) and fine spicules which can penetrate the skin, thus producing the lesions. There is evidence that the toxin and spicules are strongly sensitizing. Figure 6 shows some localized effects of sponge injuries.

Treatment consists of washing the affected area with warm soapy sea water, immersion in vinegar for 30 minutes, and then applying a cream like "Itch Balm Plus". Cyproheptadine HCl may be of value. Hot water should be avoided.

4

CNIDARIA

A. Introduction

The Cnidaria or Coelenterata are the lowest form of animal life having cells organized into different layers: ectoderm, endoderm and a noncellular layer, the mesoglea, which is a gastrovascular cavity opening only through its mouth, and is the only basic internal organ. There are no well-defined circulatory, respiratory, nervous or excretory systems. The animal is typified by radial symmetry and tentacles surrounding the mouth that are used in the capture of the prey and leading to its subsequent digestion. There are approximately 9,000 species and they vary from nearly microscopic, as in the case of some hydromedusae, to over a meter in diameter. They are found in almost all oceans and, in most cases, prefer the warmer and usually shallower waters.

The cnidarians are generally divided into three classes: Hydrozoa, Scyphozoa and Anthozoa. The Hydrozoa include the hydrozoan medusae or small jellyfishes (Fig. 7); the hydroids with branched or simple polyps, some having budded medusae, and which are usually found growing as tufts on rocks, pilings or sometimes on seaweed (Fig. 8); the hydroid or stony corals (Fig. 9); and the Siphonophora, the polymorphic, free-swimming or floating medusoids of which the Portuguese man-of-war, *Physalia* sp., is the best known (Fig. 10). The Scyphozoa, true medusae or jellyfishes, are typified by a body, umbrella, or bell, which is usually convex above and concave below. They lack a velum and stomadaeum, and the polyps may be either reduced or absent (Fig. 11). The Cubomedusae or sea wasps are the most dangerous of all the cnidarians, particularly the Indo/Australian *Chironex fleckeri* (Fig. 12) and *Chiropsalmus* sp. (Fig. 13). The common *Pelagia noctiluca* (Fig. 14a), among others (Fig. 14b), and *Chrysaora* spp. (Fig. 15) are also of considerable clinical importance. The class Anthozoa contains the corals, sea anemones and alcyonarians. The corals have calcareous skeletons and are reef builders (Fig. 16). The anemones are sedentary, flower-like structures (Fig. 17). The alcyonarians include the stony, soft, horny and black corals, as well as colonial sea pens and sea pansies, all of which lack a medusa stage. The zoanthids are best known because of the toxin palytoxin (Fig. 18). All known Anthozoa are marine creatures.

Table 8 shows some cnidarians of particular importance because of their stingings on man or their unusual toxicological properties. Some marine animals: protozoa, sponges, copepods, crabs, nudibranchs, and fishes can invade the tentacles of sea anemones without being stung. Symbiosis between these forms has been attributed to a general inhibitory effect in the anemone, to some secretion of the fish or anemone, or to some other factor. Lubbock (1979) has found in working with the clown fish, *Amphiprium clarkii*, and the anemone *Stichodactyle haddoni*, that the fish achieves its protection by means of an external mucous layer that is three to four times thicker than that found in other fishes. This layer is chiefly glycoprotein in nature, containing neutral polysaccharides. He demonstrated that the layer does not contain specific nematocyst inhibitors or excitatory substances but that it is relatively inert, causing it to be quite different from the stimulatory nature of the mucus of other fishes.

Cnidarians are of particular importance to man because of the stinging qualities of their nematocysts. The nematocysts are characteristic of the phylum Cnidaria, and "it is difficult to think of a species where they are not essential to the animal's habit of life" (Robson, 1972). Nematocysts had been classified on the basis of their structure, function and taxonomy, but until Weill (1934) proposed his elaborate nomenclature for these structures there was little common agreement on form. Weill described 17 categories of

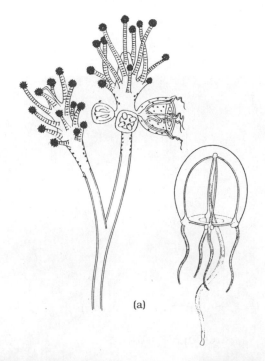

(a)

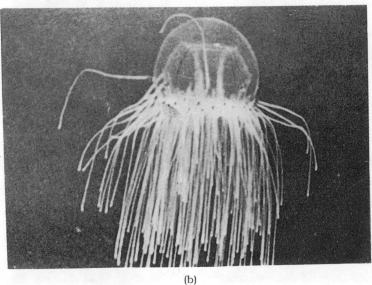

(b)

Figure 7. *Sarsia tubulosa*. (a) A common Medusa shown in its hydroid and medusa forms. (From Halstead), (b) *Gonionemus vertens*. The genus was once involved in severe stinging to 37 people swimming off Vladivostok. Six persons were said to have died. (Saunders/Russell).

Figure 8a. *Lytocarpus philippinus*, a common stinging hydroid of the Indian Ocean. (Courtesy: K. Gillett).

Figure 8b. *Lytocarpus philippinus*, a common stinging hydroid of the Indian Ocean. (Courtesy: Australian Museum).

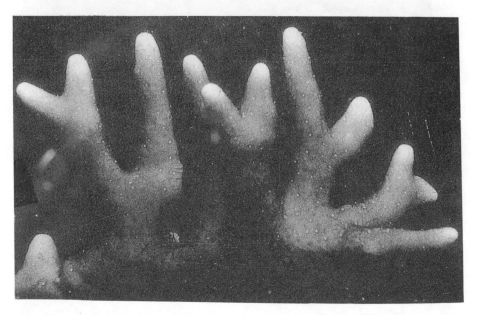

Figure 9a. *Millepora alcicornis*. A "false coral" common in the Indian Ocean as well as elsewhere. (Courtesy: R. Russo).

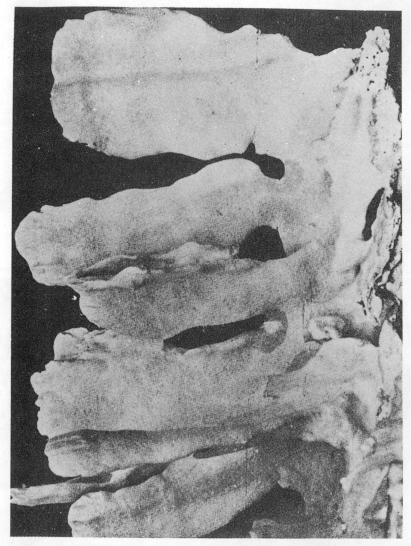

Figure 9b. *Millepora complanata*, another stinging "false coral" of the Indian Ocean. (From Boschma).

a
b

Figure 10. *Physalia physalis.* (a) F.E. Russell, (b) Specimen with captured fish in tentacles, Saunders/Russell)

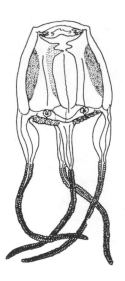
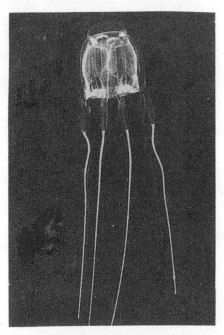

a
b

Figure 11. *Carybdea rastoni.* A small but common Indo-Pacific box jelly. Average length of bell about 2.5 cm with 5 cm long tentacles. (a) F.E. Russell, (b) Free swimming specimen. (Courtesy: R.V. Southcott)

(b)

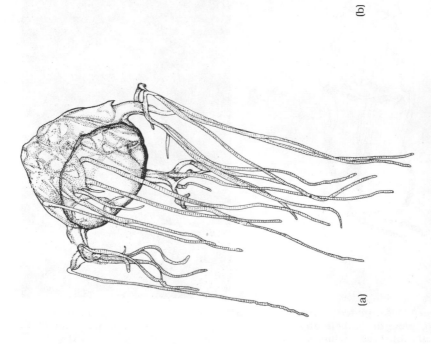

(a)

Figure 12. Chironex fleckeri. The sea wasp is the most dangerous of all the venomous marine animals. (a) F.E. Russell, (b) Courtesy: R. Endean

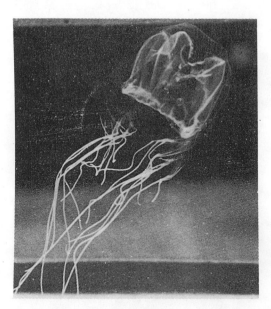

Figure 13. *Chiropsalmus quadrigatus.* (Courtesy: J.H. Barnes)

nematocysts and, while these have been modified with the
passing of time, in general his approach still offers a basis
for common communication. The interested reader is referred to the works of Hyman (1940), F.S. Russell (1953),
Hand (1961) and Werner (1965) for a more detailed discussion of nematocyst forms. Nematocysts have been classified
on the basis of their functional properties as *volent*, where
the tube end is closed; *penetrant*, where the tube end is
open; and *glutinant*, where the tube end is open and sticky.
The volent type is unarmed: its threads, when discharged,
wrap around and entangle the offending animal. The penetrant type is armed with spiralling rows of spines which
serve to anchor the thread to the prey. The stylus of the
thread is capable of penetrating some epithelial tissues, and
the venom is discharged through its open end into the
wound. The piercing ability of some penetrant nematocysts
is sufficient to puncture the chitinous cuticle of some marine
animals. The glutinant type of nematocyst responds to
mechanical stimuli, and can be used by the cnidarian for
anchoring its tentacles during locomotion.

Figure 14a. *Pelagia noctiluca.* Because of its almost circumglobal distribution this is a common cause of stingings in humans. (Maretić/Russell).

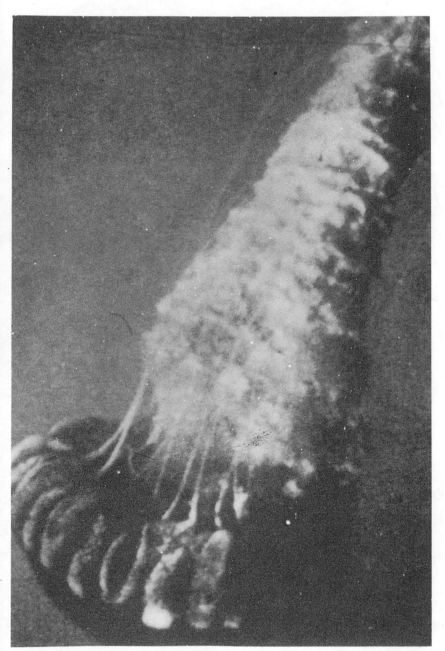

Figure 14b. *Cyanea capillata.* Thought to be the largest of jellyfishes. This specimen was 1.6 m across the belly with tentacles trailing almost 10 m. (Saunders/Russell).

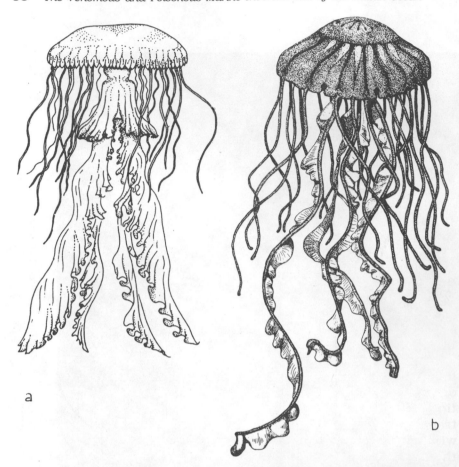

Figure 15. Two clinically important jellyfishes: (a) *Chrysaora hysoscella* (b) *Chrysaora quinquecirrha*. (F.E. Russell).

B. Venom Apparatus

The stinging unit of cnidarians is the nematocyst, a cap-sulated organelle varying in size from 4 to 225 μm, containing an operculum, a long coiled tube or hollow thread, matrix and the venom (Fig. 19). The nematocyst is formed as a "metaplasmic organelle" within a differentiated cnidocyte, the cnidoblast. At maturity these cnidoblasts are distributed throughout the epidermis, except on the basal disc. They are particularly abundant on the tentacles and are used as both offensive and defensive weapons, as well as for anchor-age. The cnidoblasts are produced at a distance from their

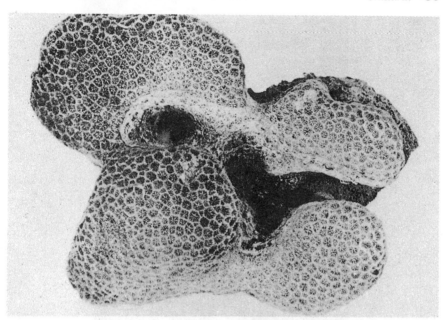

Figure 16. *Goniopora* sp. The stony corals are infrequently implicated in stingings on man. (Courtesy: B.W. Halstead).

final resting site in the epithelium; none are thought to originate in the tentacles. They migrate to their final location in the ectoderm by amoeboid activity and passive transport. The cell adjusts itself to a superficial position with that part containing the nematocyst so directed that the thread can be discharged into the ambient medium.

Extending from the free surface of the apex of the nematocyst is the cnidocil apparatus. This is a cilium-like structure, varying in length in different species. In *Physalia physalis* it is 2.5 μm. It is composed of closely packed microtubules, a basal plate and a striated rootlet. The cnidocil apparatus is thought to be the sole receptor and transducer for the discharge of the nematocyst. The cnidocil in hydrozoans, the ciliary-cone in anthozoans, and the flagellum-stereo-ciliary complex in scyphozoans are thought to be homologous sensory receptors for nematocyst discharge (Cormier and Hessinger, 1980a). In the sea anemone, activation of specific chemoreceptors by prey causes the vibration-sensitive ciliary cone to adjust to the frequency of the prey's swimming movements. This appears to be accomplished by chemoreceptor-induced changes in the lengths of

a

b

Fig. 17. *Anemonia sulcata.* (a) Common stinging anemone in many parts of the world, (b) *Actinia equina*, approximately 7 cm in diameter. (Courtesy: P.G. Giacomelli).

Figure 18. *Rhodactis howesi.* This anemone is occasionally involved in poisonings following ingestion and has been reported to have caused fatalities. (Courtesy: E.J. Martin).

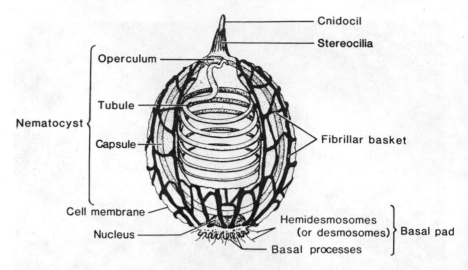

Figure 19. Diagrammatic drawing of a nematocyst from *Physalia physalis*. (From F.E. Russell)

the stereocilia that surround the cilium (Watson and Hessinger, 1991).

The question of whether or not a single or multiple mechanism is involved in the transmission of a stimulus from the cnidocil to the cell in all species of cnidarians is not fully understood. The initiating stimulus may be either mechanical or chemical, but it is probably a combination of both. In the laboratory it can be electrical. But whether or not the cnidocil transmits the message to the fibrillar collar, which exerts pressure on the capsule, deforming it and causing the operculum to be dislodged, as suggested by Cormier and Hessinger (1980a) for *P. physalis* or whether the cnidocil transmits the stimulus to a neurite or neurosecretory cell at the base of the cnidoblast, as first described by Lentz and Barnett (1965) for *Hydra*, is not easily answered. Indeed, the matter of nematocyst discharge may depend on several factors, as well as being more complex than generally believed, and it probably varies with different species of cnidarians.

In the nematocyst the coiled tubule varies in length from 50 µm to over 1 mm, depending on the species of cnidarian. Nematocyst discharge is said to be one of the fastest cellular processes known (Holstein and Tardent, 1984). When discharged, the operculum is released and the everted

Table 8. Some venomous Cnidarians.

Class	Family	Genus and species
HYDROZOA		
Medusae	Corynidae	*Sarsia tubulosa* (Sars)
	Pandeidae	*Leuckartiara gardineri* Browne
	Geryonidae	*Liriope tetraphylla* (Chamisso & Eysenhardt)
	Olindiadidae	*Gonionemus vertens* (Agassiz)
		Olindias sambaquiensis Müller
		Olindias singularis Browne
		Olindioides formosa Goto
	Pennariidae	*Pennaria tiarella* (Ayres)
Hydrozoa	Halecidae	*Halecium beani* (Johnston)
		Halecium tenellum
	Plumulariidae	*Aglaophenia cupressina* Lamouroux
		Lytocarpus pennarius (Linnaeus)
		Lytocarpus philippinus (Kirchenpauer)
		Lytocarpus phoeniceus (Busk)
Millepore corals	Milleporidae	*Millepora alcicornis* Linnaeus
		Millepora complanata Lamarck
		Millepora dichotoma Forskål
		Millepora platyphylla Hemprich & Ehrenberg
		Millepora tenera
Siphonophores	Physaliidae	*Physalia physalis* Linnaeus
		Physalia utriculus (La Martinière)
	Rhizophysidae	*Rhizophysa eysenhardti* Gegenbaur
		Rhizophysa filiformia (Forskål)
SCYPHOZOA		
Jellyfishes	Carybdeidae	*Carybdea alata* Reynaud
		Carybdea marsupialis (Linnaeus)
		Carybdea rastoni Haacke
		Tamoya gargantua Haeckel
		Tamoya haplonema Müller
	Cassiopeidae	*Cassiopea xamachana* R.P. Bigelow
	Catostylidae	*Acromitoides purpurus* (Mayer)
		Acromitoides (Acromitus) rabanchatu
		Catostylus mosaicus (Quoy & Gaimard)
	Chirodropidae	*Chirodropus gorilla* Haeckel
	(Cubomedusae)	*Chironex fleckeri* Southcott
		Chiropsalmus buitendijki Horst
		Chiropsalmus quadrigatus Haeckel
		Chiropsalmus quadrumanus (Müller)
		Carukia barnesi
	Cyaneidae	*Cyanea capillata* (Linnaeus)

(*contd.*)

(*Table 8 contd.*)

Class	Family	Genus and species
		Cyanea ferruginea Eschscholtz
		Cyanea lamarcki Peron & Lesueur
		Cyanea nozaki Kishinouye
		Cyanea purpurea Kishinouye
	Lobonematidae	*Lobonema mayeri* Light
		Lobonema smithi Mayer
	Nausithoidae	*Nausithoë punctata* Kölliker
	Pelagiidae	*Chrysaora helvola* Brandt
		Chrysaora hysoscella (Linnaeus)
		Chrysaora melanaster Brandt
		Chrysaora quinquecirrha (Desor)
		Pelagia colorata Russell
		Pelagia noctiluca (Forskål)
		Sanderia malayensis Gotte
	Rhizostomatidae	*Rhizostoma pulmo* (Macri)
	Ulmaridae	*Aurelia aurita* (Linnaeus)
ANTHOZOA		
Corals	Acroporidae	*Acropora palmata* (Lamarck)
		Acropora rosea
		Astreopora sp.
Anemones	Poritidae	*Goniopora* sp.
	Actiniidae	*Actinia cari*
		Actinia equina Linnaeus
		Actinia tenebrosa
		Anemonia indicus Parulekar
		Anemonia sulcata (Pennant)
		Anthopleura midori
		Anthopleura xanthogrammica (Brandt)
		Bunodactis elegantissima (Brandt)
		Bunodactis niobarica Carlgren
		Condylactis gigantea (Weinland)
		Physobrachia douglasi Kent
	Actinodendronidae	*Actinodendron plumosum* Haddon
	Actinodiscidae	*Rhodactis howesi* Saville Kent (poisonous)
	Agariciidae	*Payona obtusata* Crossland
	Aiptaisiidae	*Aiptasia pallida* (Verrill)
	Aliciidae	*Alicia costae* (Panceri)
		Lebrunia danae (Duchassaing & Michelotti)
	Alcyonidae	*Lophogorgia rigida*
		Sinularia abrupta Tixier-Durivault
		Sarcophyton glaucum (Quoy & Gaimard)
	Diadumenidae	*Diadumene cincta* Stephenson
	Hormathiidae	*Adamsia palliata* (Bohadsch)
		Calliactis parasitica (Couch)

(*contd.*)

(*Table 8 contd.*)

Class	Family	Genus and species
	Isopheliidae	*Telmatactis vermiformis* (Haddon)
	Metridiidae	*Metridium senile* (Verrill)
	Sagartiidae	*Corynactis australis* Haddon & Duerden
		Sagartia elegans (Dalyell)
		Sagartia longa (Verrill)
	Stoichactiidae	*Heteroactis (Radianthus) microdactylus*
		Heteroactis paumotensis (Dana)
		Stichodactyla (Stoichactis) helianthus
		Stichodactyla gigantheum Förskal
		Stichodactyla kenti (Haddon and Shackleton)
	Zoanthidae	*Palythoa tuberculosa* (Esper)
		Palythoa toxica Walsh & Bowers
		Palythoa caribbea Duchassaing
		Palythoa mammilosa Lamouroux
		Palythoa vestitus (Verrill)

tubule explodes, remaining attached at the original site of the operculum. The thread is continuous with the capsule wall, while the capsule itself is secured to a fibrillar network of microfilaments and microtubules that extend from the apex of the cell to the basal attachment of the acellular mesoglea. The structure of the tubule varies with different species. In *P. physalis* it has three spirally arranged rows of barbs that curve toward its base (Fig. 20). The open end becomes attached in the prey or victim and the venom is ejected. The manner in which the venom is ejected is not known.

Nematocyst discharge and the localized process by which these cells respond to stimuli, whether chemical, mechanical or electrical have been the object of extensive study (Parker and Van Alstyne, 1932, Weill 1934; Pantin, 1942; Yanagita, 1959a,b; Mackie, 1960; Lentz, 1966; Blanquet, 1972; Robson, 1972; Yanagita, 1973; Mariscal, 1974; Lubbock, 1979; Cormier and Hessinger, 1980b; Holstein and Tardent, 1984; Thorington and Hessinger, 1988; Watson and Hessinger, 1989a). It is not within the scope of this presentation to discuss the various theories for cell discharge. There is evidence, however, to indicate that some variations exist in the mechanisms leading up to discharge and that these, for the most part, may be related to the kind of

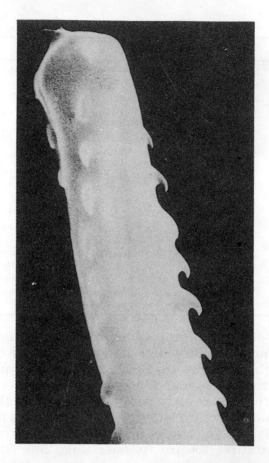

Figure 20. Discharged everted tubule from *Physalia* showing three rows of basally directed barbs. (Courtesy: D.A. Hessinger)

cnidarian involved. Contraction of the capsule, osmotic changes within the capsule, swelling of the capsular contents, dilation of the thread, and several other possible mechanisms have been suggested as events leading up to displacement of the operculum and explosion of the thread. Most recent studies would seem to indicate that the more probable mechanisms involve: (1) changes in osmotic pressure within the capsule (the osmotic hypothesis), (2) the constant pressure hypothesis, and (3) the contractile hypothesis. Cormier and Hessinger (1980b) believe that in *Physalia* the osmotic hypothesis is not applicable and suggest that the stimulus is received by the cnidocil apparatus which transmits it to the fibrillar collar with its muscle-like network of connecting microfilaments. The collar, in turn, contracts to deform the capsule, causing the operculum to be dislodged. The tubule is then discharged by the release of forces in the tightly folded inverted tubule and by hydraulic pressure. The discharging thread penetrates the prey or victim and the hooked barbs secure it, decreasing the chance of escape (Fig. 21). As more tentacles come into contact with the prey, more nematocysts are discharged (Fig. 22). Figure 23 shows a tubule within a nematocyst from *P. physalis*, within which Cormier and Hessinger (1980b) believe the venom is contained. Table 9 lists the cnidarians in Indian waters thought to be venomous.

C. Chemistry and Toxicology

Many of the earlier studies on cnidarian venoms, relating to their chemical and toxicological properties, were carried out with crude saline or water extracts prepared from the

Figure 21. Fish captured by tentacles of *Physalia physalis*. Eventually, additional tentacles were brought into play, the fish was paralyzed, retracted into the mouth and then to the gastrovascular cavity. (Saunders/Russell)

Figure 22. Nematocysts of the tentacles of *Physalia* firing following electrical stimulation. (Saunders/Russell)

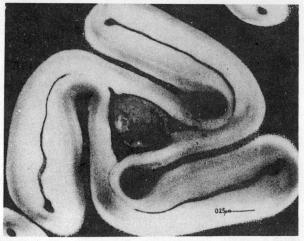

Figure 23. Undischarged tubule within a nematocyst showing venom canal. (Courtesy: D.A. Hessinger)

Table 9. Venomous cnidarians from Indian waters.

Class	Genus and Species
HYDROZOA	
Medusae	*Gonionemus vertens*
	Liriope tetraphylla
	Olindias singularis
	Pennaria sp.
	Sarsia sp.
Hydrozoa	*Aglaophenia cupressina*
	Halecium tenellum
	Lytocarpus pennarius
	Lytocarpus philippinus
	Lytocarpus phoeniceus
Millepore corals	*Millepora complanata*
	Millepora dichotoma
Siphonphores	*Diphyes chamissonis*
	Physalia physalis
	Physalia utriculus
	Rhizophysa eysenhardti
	Rhizophysa filiforma
SCYPHOZOA	
Jellyfishes	*Acromitoides purpureus*
	Acromitoides rabanchatu
	Aurelia aurita
	Carybdea alata
	Carybdea rastoni
	Catostylus mosaicus
	Chirodropus gorilla
	Chironex fleckeri
	Chiropsalmus buitendijki
	Chiropsalmus quadrigatus
	Chiropsalmus quadrumanus
	Crambionella stahlmanni
	Chrysaora helvola
	Chrysaora hysoscella
	Chrysaora quinquecirrha
	Cyanea purpurea
	Cyanea nozaki
	Lobonema mayeri
	Lobonema smithi
	Nausithoë punctata
	Pelagia noctiluca
	Rhizostoma sp.
	Sanderia malayensis
	Tamoya gargantua
	Tamoya sp.
ANTHOZOA	
Anemones	*Anemonia indicus*
	Anthopleura midori
	Bunodactis nicobarica

(contd.)

(*Table 9 contd.*)

Class	Genus and Species
	Discosoma sp.
	Metridium senile
	Parazoanthus sp.
	Physobrachia douglasi
	Rhodactis howesi
	Sinularia sp.
	Stichodactyla (Stoichiactis) giganteum
	Zoanthus sp.
Corals	*Acropora corymbasa*
	Acropora humilis
	Acropora nobilis
	Fungia sp.
	Goniastrea rectiformis
	Goniopora nigra
	Lobophytum densum
	Montipora divaricata
	Montipora foliosa
	Pavona varians
	Pocillopora eydouxi
	Porites lutea

whole animal or from one of several of its parts. The findings from these studies varied considerably and the reports on these experiments reflect the uncertainty of the nature of the toxin(s). It is apparent that many of these early workers were studying the normal constituents of the animal's tissues, several of which were found in tissues of other lower marine animals. When these substances were injected into higher animals they often produced deleterious reactions. These reactions then became aligned with clinical findings and, unfortunately, led to misunderstandings and questionable therapeutic advice (Russell, 1965a). As noted by Lane (1960), such substances as "thalassin," "congestin" and the "Cyanea principle" were probably derived from tentacular tissues rather than venom-bearing nematocysts. The earlier works on the chemical and pharmacological properties of cnidarian extracts have been reviewed by Phisalix (1922), and subsequently by Halstead (1965–70). With the works of Phillips (1956), Lenhoff *et al.* (1957), Lane and Dodge (1958), Welsh (1960), Yanagita (1959), and Barnes (1967) among others, the modern era of the study of coelenterate toxins became established. With the techniques of isolating nematocyst toxins, as proposed by Lane and Dodge

(1958), Lane (1960), Barnes (1967), Endean *et al.* (1969), Wittle *et al.* (1974), Bernheimer and Lai (1985), Kem (1973), Norton *et al.* (1990) and Senček and Maček (1990), our knowledge of the chemical nature of cnidarian toxins and their pharmacological effects has become more detailed.

Although most of the venoms of cnidarians so far studied appear to be either polypeptides or a form of protein, their pharmacology is still somewhat confused, particularly with respect to the physiopharmacological relationships between the various fractions isolated by different techniques. For instance, it is unclear whether or not some of the differences in the biological activities of the various toxins are a result of the chemical techniques used in obtaining them or are due to differences in the preparations used to define the pharmacological property. In addition, one wonders if the proteinase inhibitor of a toxin has its significance, physiologically or pharmacologically in this activity, or if this property is a subunit of a more basic property in the armament of the venom, and one for which our techniques are not yet geared, or are not functionally applicable. These matters are not easily resolved. In general, the toxins of cnidarian nematocysts probably serve several purposes: (1) they immobilize the prey so that the tentacles may deliver the prey to the oral cavity, (2) the cytolytic components begin the digestion of the prey, and (3) they deter potential predators.

Because of the numerous coelenterates having nematocyst-bearing toxins and differing systematically, it seems wisest to consider the chemistry and pharmacology of these poisons on the basis of the animal's family and class.

1. *Hydrozoa, Medusae, Olindiadidae*

Using thin-layer chromatography, Sephadex G-200 and bioassay with rabbit platelets, it was found that the isolated glycoproteins of the medusa *Gonionemus vertens* had eight main fractions, with that of 30 ± 5 kD being the most toxic, although all fractions were active. Lipids and glycoproteins from the tentacles were isolated and separated by thin-layer chromatography (TLC) and platelet activating factor (PAF)-corresponding TLC-zones and were tested on rabbit platelet-enriched plasma. Studies showed the presence of

PAF in the lipids, and all glycoprotein fractions contained PAF (Berdyshev *et al.*, 1991).

2. *Hydrozoa, Millepore Corals, Milleporidae*

Modern fire coral toxicology began with the works of Wittle *et al.* (1971) and Middlebrook *et al.* (1971). The former investigators studied nematocyst toxin from *Millepora alcicornis*, which they obtained by exposing the coral to a pH 5.8–7.0 phosphate buffer, precipitating with 80 percent ammonium sulfate, and then passing the product through DEAE-cellulose, Sephadex G-100 and hydroxyapatite. The partially purified toxin gave three protein bands on acrylamide disc electrophoresis. From Sephadex G-100 studies, a molecular weight of approximately 100,000 was suggested. The toxin was relatively stable and its i.v. LD_{50} in mice was found to be 0.04 mg/kg body weight. Subsequently, using undischarged nematocysts from *M. tenera*, this group found that the toxin obtained from the undischarged nematocysts had a similar LD_{50} and concluded that the toxin from the undischarged nematocysts was the same as that obtained from nematocysts discharged by exposure to the phosphate buffer. The toxin had hemolytic and dermonecrotic activities, and was antigenic with cross protection against *M. alcicornis* toxin (Wittle and Wheeler, 1974). Middlebrook *et al.* (1971) found that they could obtain an electrophoretically pure toxin from the fire coral *Millepora dichotoma* by a simple pass through DEAE-cellulose following the same initial extraction procedures. The LD_{50} was 0.038 mg/kg body weight, quite similar to that found for *M. alcicornis*. The signs in mice were also similar.

3. *Hydrozoa, Medusae, Siphonophora*

Lane and Dodge (1958) and Lane (1960) obtained an active preparation from *Physalia physalis* by isolating nematocysts from the tentacles by autolysis at 4°C for 24 to 48 hours, and then by passing a seawater solution of the digested tentacles through a graded screen to isolate the undischarged nematocysts. These were washed repeatedly with sea water and allowed to settle overnight. The supernatant was decanted and the residue, composed chiefly of nematocysts, was centrifuged at 300–400 rpm for 15 to 30 minutes and then resuspended in sea water. The suspension was thought to be almost completely free of tentacular tissue and contained approximately 55 million undischarged nematocysts/g wet weight (from 3.8 liters of tentacles). The contents of the nematocysts were freed by homogenization, the homogenate centrifuged to separate the capsules and capsular fragments from the capsular contents, and the supernatant retained for subsequent chemical and pharmacological studies. Using this technique, the authors found the supernatant to be a highly labile protein complex, rich in glutamic acid, having an approximate intraperitoneal lethal dose in mice of 0.037 ml/kg body weight of a preparation containing 0.02 percent total nitrogen.

This toxin produced paralysis in fishes, frogs and mice. Animals killed following stingings by *Physalia* exhibited marked pulmonary edema, right cardiac dilation, with venous congestion of the larger vessels of the chest and portal circulations. It was suggested that the toxin affected the respiratory centers before producing changes in the voluntary muscles, and that it altered the permeability of the capillary wall but did not produce hemolysis. It also caused changes in the isolated heart of the clam, which the authors thought resembled those provoked by acetylcholine.

Lane (1961) subjected lyophilized "crude" extracts of *Physalia* nematocysts to one-dimensional chromatography and obtained nine spots, four of which accounted for 95 percent of the total lethal activity for the crab *Uca pugilator*. By paper electrophoresis, he separated the same extracts into four fractions, three of which contained the total lethality, the principal lethal portion being in two of the fractions. The crude toxin was lethal to mice at 1.7 mg/kg body weight. He suggested that *Physalia* toxin is a relatively simple substance, consisting of only a few toxic peptides that are synthesized by gastrodermal cells and which pass through the mesoglea and then into the nematocyst during the morphogenesis of this structure. In a further study with paper electrophoresis, *Physalia* toxin was separated into four fractions, two of which accounted for 95 percent of the toxicity, and in polyacrylamide gel electrophoresis the toxin was resolved into 8 to 10 fractions. The crude preparation had protease activity (Hines and Lane, 1962). Subsequently, phospholipases A and B were detected in the crude venom, but were not thought to be related to lethality (Stillway and Lane, 1971).

Various studies on crabs, rats, dogs, and a nerve-muscle preparation of the frog indicated that the toxin produced changes in the Na-K pump, resulting in depolarization of the cell membranes (Lane, 1968; Larsen and Lane, 1970). It produced a marked change in the normal pattern of pressure variations in the hemocoel and electrocardiogram of the land crab *Cardisoma guanhumi* (Lane and Larsen, 1965). In the rat, where the LD_{50} was approximately 100 µg/ kg body weight, low doses of the toxin caused an increase in the Q-T interval, a decrease in the P-R interval and P wave inversion in the electrocardiogram. Large doses produced marked ECG changes leading to cardiac failure (Larsen and Lane, 1966). These authors postulated that, if *Physalia* toxin causes a general sustained depolarization of the postsynaptic membrane, this might account for the differences in the responses of the mammalian (myogenic) and crustacean (neurogenic) hearts to the toxin.

Néeman *et al.* (1980a) demonstrated that the nematocyst venom contained a DNase which had a non-specific endoleolytic action. The enzyme had a molecular weight of approximately 75,000, was thermolabile and its activity was stimulated by 80 mM NaCl or 10 mM MgCl. Tamkun and Hessinger (1981) obtained a hemolytic protein from *P. physalis*, using nematocyst isolation methods as suggested by Lane and Dodge (1958). This protein physalitoxin, was also lethal to mice at the 0.20 mg protein/kg body weight level, while the LD_{50} for the crude venom was 0.14 mg/kg. A molecular weight of 212,000 was calculated. The sedimentation coefficient was 7.85, and the structure was rod-like in shape with a calculated axial ratio of approximately 1:10. The authors suggested that the toxin was composed of three subunits of unequal size, each of which is glycosylated. Physalitoxin was found to be about 28 percent of the total nematocyst venom protein. Its carbohydrate content was 10.6 percent and it represented the major glycoprotein of the crude venom. This hemolytic and lethal toxin was inactivated by concanavalin A. The inactivation was blocked in the presence of α-methyl-mannoside. This would seem to indicate that the inactivation by concanavalin A is probably caused by an interaction with specific saccharides on the hemolysin. The inactivation was temperature-dependent above 12°C.

Mas *et al.* (1989) investigated the effects of a high molecular weight toxin (P_3) from *Physalia physalis* on the glutamate evoked potentials in the snail and crayfish neuromuscular junction. P_3 produced a reversible block of the excitatory junctional potentials and on glutamate potentials in the crayfish at concentrations of 6 nM–60 nM. The action was mainly on excitatory junctional potentials. The crude venom from *Physalia* species from the Arabian Sea has been shown to have somewhat similar pharmacological properties to the venom of *Physalia* species from other areas of the world, although apparently is less toxic. The LD_{50} of specimens from this region was 4 mg/kg body weight; it showed hemolytic and dermonecrotic activities at protein doses of 0.5 mg/ml for the former activity and 0.1–0.35 mg/ml for the latter. Long-term freezing and two to three freeze-thaw cycles caused a gradual loss of lethal and hemolytic activities, but the dermonecrotic property was little affected (Alam and Qasim, 1991).

4. *Scyphozoa, Jellyfishes, Cubomedusae*

The initial studies on the chemistry of *Chironex fleckeri*, commonly known as the sea wasp, were carried out by Southcott and Kingston in 1959. It remained, however, for Barnes (1967a) to isolate a crude nematocyst toxin from this cubomedusid. This was accomplished by the ingenious method of using human amnion membrane and electrically stimulating the tentacles (Fig. 24). In his initial study, Barnes found that the undiluted toxin was lethal to mice at the 0.005 ml/kg body weight level, but it was not known what this might be in dry weight, mg protein or protein nitrogen.

Using a modification of the extraction methods proposed by Phillips (1956) and by Lane and Dodge (1958), Endean *et al.* (1969) identified five types of nematocysts in the tentacles of *C. fleckeri*. They found proteins, carbohydrates, cysteine-containing compounds and 3-indolyl derivatives in all five types. Saline extracts of the contents of the nematocysts were highly toxic to prawn and fish, and were lethal to mice and rats. In mice, the intravenous LD_{50} was between-20,000 and 25,000 nematocysts, while in rats the LD_{50} was approximately 150,000 nematocysts. Extracts of

Figure 24. Equipment used by J.H. Barnes (1967a) for milking venom from living tentacles: concave human placenta membrane over domestic coffee jar, short plastic transfusion tubing with connector and clamp to flexible sidearm through a hole at the bottom of the jar, and other ancillaries. Electrical stimulation was provided by a Multi-Tone Progressive Treatment Unit.

ruptured nematocysts elicited a strong contraction in bar-nacle striated muscle and in the skeletal, respiratory and extravascular smooth muscle of the rat. The contracture was sustained for varying periods, then the muscle became paralyzed in the relaxed state. When the perfused heart of a toad and the exposed heart of a rat were subjected to the nematocyst extract, there was a progressive failure in relax-ation during succeeding cardiac cycles, and the heart became paralyzed in systole. Prior exposure of rat diaphragm mus-culature to D-tubocurarine did not modify the response to the toxin, and conduction in the toad sciatic nerve was unaffected by prolonged exposure to the toxin.

Using extracts of the tentacles partially purified by Sephadex gel filtration, Freeman and Turner (1969) per-formed a number of pharmacological tests in mammals and concluded that the extracts produced respiratory arrest which they attributed to a central origin. They also implicated deleterious cardiac changes leading to an atrioventricular block. Blood pressure and chemistry changes were consistent with a reduced circulating blood volume and hypoxia. The toxins had a non-specific lytic effect on cells and no particular differential effect on the guinea-pig diaphragm preparation. It was found that 0.1 ml of a 5000-fold dilution of the tentacle extract would kill a 20 g mouse in less than two minutes, and that the toxin was hemolytic. The toxic fraction was nondialyzable and was eluted over a range from bovine serum albumin to a molecular weight of 8000.

Crone and Keen (1969), using tentacle extracts as described by Freeman and Turner (1969), obtained two toxic proteins by ion exchange chromatography on carboxymethyl and DEAE cellulose, and by exclusion chromatography on Sephadex G-75 and G-200. The hemolytic activity was related to a protein component with a molecular weight of approximately 70,000. The second toxin had a molecular weight of about 150,000. While both components had cardiotoxic activity, the larger fraction had considerably more than the smaller one. The studies indicated that there was only one hemolytic fraction in the extract. This hemolysin was labile at room temperature and there was a non-linear relationship between the rate of hemolysis and dilution of the sample that was dependent on temperature and pH. Activity was inhibited by sucrose and plasma, and accelerated by benzene. There was a correlation between the hemolysis titer and the LD_{50} (Keen and Crone, 1969a).

The cardiovascular effects of two *Chironex* tentacle extracts were studied by Freeman and Turner (1971), who found that both the fractions isolated previously produced an initial increase in systemic arterial pressure followed by a fall in pressure, bradycardia and cardiac arrhythmia. In the perfused guinea-pig heart, both toxins caused a reduction in rate, amplitude of contraction and coronary flow. The authors concluded that the cardiovascular effects were due to "direct vasoconstriction, cardiotoxicity, a baroreceptor stimulation and possible depression of the vasomotor center."

From the various studies on *C. fleckeri*, it became apparent that the toxic material present in the nematocysts and that present in the tentacles devoid of nematocysts was lethal to mice and rats but that these materials possessed markedly different biological activities. At the time it was concluded that:

Cognizance should be taken of the presence in *C. fleckeri* tentacles of toxic material which is not localized in nematocysts when studies are contemplated of the toxic material normally injected via nematocysts into prospective prey or into humans who come into contact with the tentacles of the jellyfish. Moreover, the walls of the nematocysts of *C. fleckeri* are extremely resistant to mechanical abrasion and prolonged grinding in an electrically driven microgrinder is necessary to bring about rupture of the majority of nematocysts (Endean *et al.*, 1969). Hence it is unlikely that mincing or a brief homogenization of the tentacles of *C. fleckeri* would rupture a significant number of nematocysts. Because of the method of preparation of toxic extracts adopted by Freeman and Turner (1969), it is possible that these authors studied the activity of toxic material located in the tentacles of *C. fleckeri* but which may not necessarily be present in the nematocysts.... Because both toxic materials have hemolytic properties, it is not immediately apparent which material was studied by Keen and Crone (1969a) and by Crone and Keen (1969), but the method of preparation adopted by the authors would suggest that they may have studied a toxic material from the tentacles which is not localized in the nematocysts. On the other hand, toxic material prepared in a similar fashion by Keen and Crone (1969b) showed dermonecrotic activity which is also shown by toxic material localized in the nematocysts but not by toxic material from tentacles devoid of nematocysts. Also, Freeman and Turner (1969) and Keen and Crone (1969) state that the pharmacological properties of the extracts of whole tentacle which they had prepared were identical with those obtained from material collected by discharge of nematocysts through a human amniotic membrane (Russell, 1971a).

In 1971, Endean and Noble sought to clarify some of the differences in the findings for this venom. Using methods previously perfected by Endean *et al.* (1969) and Endean and Henderson (1969), they separated material within the nematocysts from the residual tentacular material. The two products were studied by injection into mice and rats, on the barnacle muscle preparation, the rat phrenic nerve-diaphragm preparation, the toad sciatic nerve-gastrocnemius and sciatic nerve conduction preparations, the rat ileum and heart preparations and several other isolated tissues.

It was found that following removal of the nematocysts from the tentacles, the remaining products possessed quite different biological activities from that extracted from the nematocysts. These findings have been summarized in Table 10. On the basis of these results the authors concluded:

> It is possible that the pharmacological properties of toxic material present inside the nematocysts of *C. fleckeri* are modified during discharge, during subsequent standing or storage, or during the extraction procedures adopted by different authors who have studied toxic material from this jellyfish. The toxic material found in tentacles devoid of nematocysts could represent material which has been discharged previously from nematocysts and adsorbed on the tentacles and which had undergone a change in its pharmacological properties. However, the weight of evidence indicates that the substance of the tentacle normally contains toxic material different from that present in nematocysts (Endean and Noble, 1971).

Although the work of Endean and Noble (1971) indicates certain physiopharmacological differences, it is difficult to define these precisely in the absence of a standard tissue weight or solution. Comparing a solution containing 25,000 nematocysts with a tentacular solution of 0–10 ml leaves room for speculation.

Also, some of the differences demonstrated by various investigators for *Chironex* toxins (as well as other cnidarian

Table 10. Differences in pharmacological activities with extracts from tentacles and nematocysts of *Chironex fleckeri.*

Preparation	Tentacle extract	Nematocyst extract
Muscle, barnacle	Unaffected	Affected
Heart, rat	Early hypotension	Early hypertension
Nerve, rat	Block	No block
Skin, rat	Very slight necrosis	Necrosis
Blood cells, rat	Hemolysis	Hemolysis

toxins) might well be due to dose relationships rather than the kind of toxin employed. This seems particularly true for the rat or guinea-pig diaphragm-phrenic nerve preparations. Changes in the dose of toxins can not only cause quantitative changes but also qualitative ones. Further, when the muscle of a mammalian nerve-muscle preparation is shortened it is most difficult to determine the effect on the indirectly elicited contractions. Further, in a shortened muscle it is even difficult to determine the significance of the direct effect. One may speculate but direct comparisons are questionable. Even adding distilled water to the bath will change the resting tension of a fiber to the extent that both qualitative and quantitative alterations can occur. Certainly, a substance that is highly irritating to an entire muscle membrane presents a problem when one hopes to define its action on the nerve, or even on the activity of a single muscle fiber.

Baxter and his colleagues at the Commonwealth Serum Laboratories in Australia, using saline extracts of nematocysts obtained after the method of Barnes, or as described by Endean *et al.* (1969), demonstrated that the biologically active fractions causing lethal, hemolytic anddermonecrotizing effects were in the 10,000–30,000 molecular weight range. The activities were minimized by heat, formalin and EDTA, and were stabilized by peptone. There was thought to be no histamine-releasing action in the dermonecrotizing factor, and the hemolysin was not a phospholipase (Baxter and Marr, 1969).

In rabbits, a lethal intravenous dose of the venom caused labored and deep respirations followed within several minutes by prostration, hyperextension of the head, "spasms," and respiratory and cardiac arrest. In sheep, a similar respiratory deficit was seen, followed by unsteadiness, muscle tremors and "spasms," The head drooped to one side, prostration occurred, and the tongue was said to be paralyzed and cyanotic. The animal lay on its side kicking and gasping, eye reflexes were lost, and there was a generalized body tremor. Death occurred within a few minutes. In primates, the animal became inactive, "dull," confused, and exhibited slight ataxia and incoordination. The eyelids drooped, the mouth was open and the head "weaved." Heart rate was irregular and breathing became labored, deep and irregular.

Within a few minutes the animal collapsed, the heart rate deteriorated and cyanosis developed just prior to death. Post-mortem examination revealed marked congestion of the vessels of the lungs and pulmonary edema. The right ventricle was engorged, the kidneys and liver were congested, and the vessels of the meninges of the cerebrum were engorged (Baxter *et al.*, 1972).

Early studies on the preparation of a *Chironex fleckeri* antivenin are difficult to interpret. In 1974, Baxter and Marr found that antivenin prepared against *C. fleckeri* venom neutralized extracts of *Chiropsalmus quadrigatus* and an unidentified tropical cubomedusan *in vitro* but not those of *Physalia physalis* or *Chrysaora quinquecirrha*. Again, the data are difficult to interpret because the standardized chemical units were not used. Nevertheless, the conclusions seem valid. Subsequently, a sea wasp toxoid and an antivenin were prepared (Baxter and Marr, 1975). However, preparation of the toxoid was abandoned for it was found that there was only a remote chance of an immunized person being stung, the period of effective protection was short, and there was the possibility of sensitizing individuals (S. Sutherland, personal communication, 1981). The antivenin, however, has been found very effective for *C. fleckeri* stings (Sutherland, 1983).

The lethal and hemolytic properties of *Chironex fleckeri* have been stabilized by Comis *et al.* (1989), who found that the toxin's activities were higher when no buffers or water were used during the initial extraction, which had been a common procedure in earlier studies. When the extract was first filtered through a Sep-pak C_{18} cartridge its lethal titer was increased 16-fold, and hemolysis activity 2.6-fold. Proteolytic activity was also demonstrated. In 1993, Collins *et al.* reported on the development of monoclonal antibodies bound to a 50,000 dalton toxin associated with the hemolytic activity of the extracts of *C. fleckeri*.

5. *Scyphozoa, Jellyfishes, Pelagiidae*

Four distinct type of nematocysts have been observed in the sea nettle *Chrysaora quinquecirrha*: homotrichous anisorhiza, homotrichous isorhiza, a smaller isorhiza and, the most abundant type, the microbasic eurytele (Burnett *et al.*, 1968;

Blanquet, 1972). Using the nematocyst isolation method of Rice and Powell (1970), Blanquet (1972) found that the toxin was contained within the nematocyst and that the discharged nematocyst capsules and threads were free of toxin. The crude toxin was heat labile and toxicity was abolished following boiling for five minutes. However, the toxin could be lyophilized or frozen and stored at –20°C for at least a month without a significant loss of activity. Toxicity was lost following incubation for one hour with trypsin or after protein precipitation with 10 percent trichloroacetic acid. Further studies showed that the toxic material was associated with a protein fraction having a molecular weight >100,000 which could be separated into two major fractions. The more toxic of these two proteins was found to be rich in aspartic and glutamic acids, which comprised approximately 27 percent of the total detectable amino acid content.

Burnett and Carlton (1977) have reviewed the cardiotoxic, dermonecrotic, musculotoxic and neurotoxic properties of *C. quinquecirrha* venom. Neéman *et al.* (1980a,b) demonstrated that the toxin, separated after a modified method of Burnett *et al.* (1968), produced striking cytological changes, including nuclear alterations and dissolution of intercellular collagen. The lethal property is thought to exert its effect by altering the transport of calcium across the conduction system of the heart (Kleinhaus *et al.*, 1973). From experiments on rat and frog nerves and muscles, and the neurons of *Aplysia californica*, Warnick *et al.* (1981) concluded that the toxin appeared to induce a non-specific membrane depolarization by a sodium-dependent tetrodotoxin-insensitive mechanism which secondarily increases Ca influx; that Ca^{++} does not play a direct role in the depolarization caused by the toxin; that the action is not on voltage-sensitive Na^+ and K^+ channels; and that it is not dependent upon lipase activity. The data seem consistent, although some workers might question the authors conclusions.

Lal *et al.* (1981) isolated the collagenase from *C. quinquecirrha* and purified it 237-fold, using three sequential ion exchange chromatographic steps. It was activated by Ca^{++} and Na^+, reversibly inhibited by EDTA, uninhibited by phenylmethyl-sulfonyl fluoride but inhibited by a number of chelating agents, cystine and some metal ions. A monoclonal antibody to the lethal factor of the venom has been prepared, using the crude venom as antigen. Cloned hybridomas were selected by ELISA, using the partially purified lethal factor. Ascites fluid from the cloned hybridoma-breeding mouse showed an ELISA titer of 12,800 and neutralized 2-LD$_{50}$ of an intravenous injection of the crude venom following incubation overnight (Gaur *et al.*, 1981).

In studying the ultrastructural changes caused by the nematocyst toxin of *C. quinquecirrha* in cultured hamster ovary K-1 cells, Neeman *et al.* (1991) suggest that the nuclear changes and the dissolution of intercellular collagen were probably due to the collagenase, protease and lectin-like activities. In a rat model, Muhvich *et al.* (1991) have shown that the venom produced hepatic and renal necrosis without causing histological alterations in the brain, heart or lungs. Hepatocytes in the mid-zonal areas and renal tubular epithelium were the cell types chiefly affected. Neither hyperbaric oxygen nor verapamil had any effect on the extent of the hepatic injury, nor did the oxygen protect against the venom-induced decrease in the blood flow in the brain. The authors note that hyperbaric oxygen caused no changes in cardiac output but there was a 40 percent reduction in blood flow to the brain probably inducing ischemic cerebral dysfunction. Respiratory arrest, whether central or peripheral, or both, is generally stated as the mechanism leading to death.

6. *Anthozoa, Corals and Anemones*

Since some ciguateric fishes were known to feed on corals, early writers sometimes attributed this fish poison to the soft corals. Hashimoto and Ashida (1973), however, were unable to demonstrate ciguatoxin, the major toxin involved in ciguatera poison, as a component of the coral samples they studied. They did find several water-soluble toxic fractions in the anthozoans *Goniopora* sp., *Palythoa tuberculosa*, *Coeloseris mayeri*, *Clavularia* sp., *Acropora* sp., *Cyphastrea* sp. and *Pavona ebtusata*. *Goniopora* toxin was prepared as shown in Chart 1. It was found to be rich in aspartic acid, proline, isoleucine, leucine and lysine. The authors suggested a molecular weight of 12,000. Its intraperitoneal minimal lethal dose in mice was 0.3 mg/kg body weight. The mice exhibited hypoactivity, followed by rigor, respira-

Chart 1. Purification scheme for the toxin from *Goniopora* sp. (modified from Hashimoto, 1979).

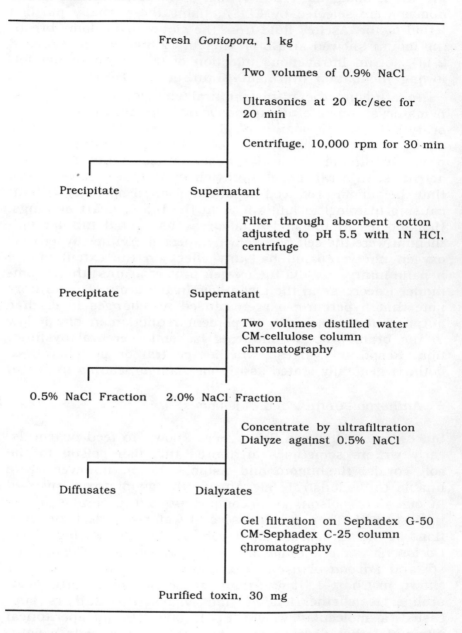

Fresh *Goniopora*, 1 kg

Two volumes of 0.9% NaCl

Ultrasonics at 20 kc/sec for 20 min

Centrifuge, 10,000 rpm for 30 min

Precipitate   Supernatant

Filter through absorbent cotton adjusted to pH 5.5 with 1N HCl, centrifuge

Precipitate   Supernatant

Two volumes distilled water CM-cellulose column chromatography

0.5% NaCl Fraction   2.0% NaCl Fraction

Concentrate by ultrafiltration Dialyze against 0.5% NaCl

Diffusates   Dialyzates

Gel filtration on Sephadex G-50 CM-Sephadex C-25 column chromatography

Purified toxin, 30 mg

tory distress, paralysis and cyanosis. Skeletal muscle rigidity just prior to death was a characteristic finding. A 0.9 percent saline extract of coral had a hemolytic effect on rabbit erythrocytes and, when purified on Sephadex G-50, followed by chromatography on CM Sephadex C-50, its hemolytic activity was increased to that of Merck saponin. It had a suspected molecular weight of 30,000. On disc electrophoresis it still showed three bands (Hashimoto, 1979).

Interest in the alcyonarians was elicited by Neeman *et al.* (1974), who observed that certain "soft corals" found in the Red Sea were not attacked by predators. It was observed that fish would die if they were kept in the same container with *Sarcophyton glaucum*. This alcyonarian was extracted with hexane in a Soxhlet over a 24-hour period. After cooling the extract, a white crystalline precipitate appeared. The crystals were isolated, dried and, after recrystallization from acetone-hexane, their melting point was found to be 133-134°C and (αD^{25} + 92°, 1.0, $CHCl_3$). NMR were taken using a solution with TMS as an internal standard. Mass spectra were done. The toxin, called sarcophine, $C_{20}H_{28}O_3$, was subsequently shown to be:

(From Erman and Neéman, 1977)

It contains three methyl groups, an additional methyl on an epoxide bearing carbon, as well as three particular downfield appearing protons. The LD_{50} in the fish *Gambusia affinis* was 3 mg/l of water. The first deleterious signs were excitation, turning and jumping, followed by prostration, sinking to the bottom and cessation of gill cover movement. In rodents, subcutaneous injections caused excitation, increased respirations, and "paralysis" followed by coma and death. When rats were fed sarcophine there was a decrease in cardiac, respiratory and motor activity. On the isolated guinea-pig ileum, 0.2–0.8 mg of the toxin/l bath solution produced a strong "anticholinesterase" response. The authors suggested that materials like sarcophine form part of the mucus secreted by the coral and act as a repellant. Erman and Neéman (1977) found that sarcophine inhibited the enzyme phosphofructokinase. Linear uncompetitive inhibition of the enzyme was thought to be due to the reaction of the sarcophine with thiol groups of the enzyme.

Anemones of the family Actiniidae are of particular interest because of their stings on humans and the chemical nature of their toxins. The research prior to 1964 has been reviewed by Halstead (1965–70). Ferlan and Lebez (1974) isolated a highly basic protein toxin from *Actinia equina* by autolyzing a suspension of nematocysts from the tentacles, passing it through a thick nylon screen to obtain the undischarged nematocysts, freezing and thawing the nematocysts, homogenizing and centrifuging the product, and then precipitating with acetone. The precipitate was chromatographed on Sephadex G-50 and an electrophoretically homogenous protein having a molecular weight of 20,000 and an isoelectric point of 12:5 was obtained. The toxin, called equinatoxin, had 147 amino acid residues (Ferlan and Lebez, 1976). In rats, equinatoxin had an intravenous LD_{50} of 33 µg/kg body weight, and was found to have hemolytic, antigenic, cardiotropic and had certain other activities (Sket *et al.*, 1974). Ferlan (1977) found that *in vitro*, equinatoxin exhibited strong lytic action on erythrocytes, and that it did not have phospholipase activity. He also studied the effects of different ions on the toxin's hemolytic properties.

Giraldi *et al.* (1976) demonstrated that a suspension of phospholipids would reduce hemolytic activity. Pencar *et al.* (1975) demonstrated that a greater concentration of the toxin was required to cause changes in the molecular order of the membrane than that required for complete hemolysis. Maček and Lebez (1981) showed that 1×10^{-10} M of the toxin caused complete hemolysis of sheep erythrocytes, while Ca^{++}, Mg^{++} and Ba^{++}, in that order, stimulated the rate of hemolysis. EDTA decreased hemolytic activity by 30 percent, while Sr^{++} and Mn^{++} had no effect. The authors concluded that the data did not clarify the role these ions play in equinatoxin-induced hemolysis. At neutral and alkaline pH it is hemolytic (Maček and Lebez, 1981), induces pulmonary edema (Lafranconi *et al.*, 1984), causes aggregation and lysis of platelets (Teng *et al.*, 1988), and has been shown to have cardiotropic activity (Ho *et al.*, 1987). EqT II is closely related to the other equinatoxin components and has distinct cytotoxic properties inhibited by membrane phosphatides, particularly sphingomyelin (Kem, 1988). It is approximately 20 kD, has an i.v. LD_{50} in mice of 35 μg/kg and has humoral as well as cellular immune responses in mice (Strazisar, personal correspondence). It has been studied in various cell cultures (Giraldi *et al.*1976) and reviewed with additional data by Batista *et al.* (1990). Equinatoxin is now known as a group of toxins, Equinatoxins I, II, III.

Senček and Maček (1990) milked individual specimens of *Actinia cari* by pressing them over collecting beakers while still alive and at 2, 15 and 35 days, then once a month for four months. The venom samples were pooled, centrifuged and applied to a Sephadex G-15 column. They obtained two lethal and hemolytic polypeptides, cardiotoxins I and II by gel and ion exchange chromatography. The molecular weight of the toxin was 19,800. The isolectric points were 9.45 (I) and 10.0 (II). The hemolytic activity was inhibited by sphingomyelin. A cytolytic toxin, metridiolysin, from the anemone *Metridium senile* was also shown to have similar hemolytic activity, inhibited by cholesterol (Bernheimer and Avigad, 1978). Like equinatoxin, hemolytic activity was restricted to a relatively narrow pH range, with an optimum at 5.6. The optimum was 8.8 and 8.5 in the presence of Ca^{++}. Equinatoxin exhibits no hemolytic activity below a pH value of 6.5.

Several other Actiniidae species have been shown to contain toxins. Norton *et al.* (1976) isolated a polypeptide, termed anthopleurin-A, from *Anthopleura xanthogrammica* that closely resembles toxin II of *Anemonia sulcata* in its amino acid sequence (Tanaka *et al.*, 1977). In mice, anthopleurin-A had an LD_{50} of 0.3–0.4 mg/kg body weight and was said to stimulate cardiac activity. A second polypeptide toxin, anthopleurin-B, was also identified in the same anemone. Norton, *et al.* (1990) have demonstrated that three proteins from *Actinia tenebrosa* have hemolytic and cardiac stimulating activities. They are basic (pI 9.4), have a molecular weight of approximately 20,000 and possess highly homologous amino acid compositions and N-terminal amino acid sequences. The terminal sequences showed homology with that of equinatoxin from *Actinia equina* and cytolysin III from *Stichodactyla helianthus*. None of the proteins had cystein or cysteine residues. They exhibited strong positive inotropic effects on guinea pig atria and had hemolytic activity.

Considerable study has been done on the chemistry of the anemone *Anemonia sulcata* and other members of the genus. In 1973, Novak *et al.* obtained a partially purified, toxic, basic polypeptide from *A. sulcata* by homogenizing the tentacles, freezing the homogenate in liquid nitrogen and then thawing it repeatedly to break down the nematocysts. The extract was fractionally precipitated with ammonium sulfate and chromatographed on DEAE-cellulose and Sephadex G-50. Its molecular weight was estimated to be 6000; it gave four bands on disc electrophoresis. Its LD_{100} in rats was 6 mg/kg body weight.

In 1975, Beress *et al.* isolated three toxic polypeptides (toxins I, II and III) from *A. sulcata* by homogenizing specimens with ethanol, heating to 60°C, and then dialyzing and exposing the extract to CM cellulose, Sephadex G-50, SP Sephadex, and Bio-Gel P-2 to separate and purify each of the toxins. Toxin I contained 45 amino acid residues, toxin II, 44 and toxin III, 24. Toxins I and II had glycine at the N-terminus, while in toxin III it was arginine. The molecular weights of all three were less than 4750. Toxins I and II had similar toxicological effects. When injected into crustaceans, fishes and mammals they produced paralysis and cardiovascular changes. When injected into crabs the toxins caused convulsions and paralysis, and were lethal at 2.0 μg/kg. They also caused paralysis and death in fishes. The intramuscular LD_{50} in the juvenile cod *Gadus morhua* was 6.9 mg/kg body weight for toxin I and 0.14 mg/kg for toxin II (Möller and Beress, 1975). Toxin III caused neurotransmitter release from rat synaptosomes, as did veratridine, but the release activity appeared to occur at different receptor structures in the membrane (Abita *et al.*, 1977). Toxin III of *A. sulcata* contains 27 amino acid residues and according to Bahraoni *et al.* (1989), has no sequence similarity with toxins I and II from the same species.

In electrophysiological studies on crayfish neuromuscular transmission, Rathmayer *et al.* (1975) demonstrated the toxicity of toxins I and II, while Alsen and Reinberg (1976) noted that they were unable to show neurotoxicity in the mouse. They had found that toxin II was far more toxic than toxin I, and that it produced a positive inotropic effect on isolated electrically driven atria of the guinea-pig at concentrations of 2×10^{-8} g/ml. At higher concentrations it caused contracture and arrhythmia. On the Lagendorf heart preparation, low concentrations enhanced the contractile force of the atrium and ventricle, while high concentrations, 1×10^{-7} g/ml, caused contracture and arrhythmia, which appeared to be limited to the atrium. The atria proved to be much more sensitive to the toxin than the ventricular muscle.

In 1964, Munro demonstrated that tentacle homogenates of the large anemone *Condylactis gigantea* had a marked paralytic effect on crustaceans. Taking advantage of this observation, Shapiro and his colleagues carried out a series of chemical and pharmacological studies on a stable acetone powder from tentacle homogenates of the species (Shapiro, 1968a,b; Shapiro and Lilleheil, 1969). The tentacles were washed, blotted, cooled, homogenized and centrifuged at 10,000 rpm for one hour. Cold acetone was added to the supernatant, the suspension filtered and washed with acetone and then dried. The uniform, fine yellow powder was stored at $-15°C$ under vacuum. It was then resuspended in cold buffer, homogenized and centrifuged at 60,000 rpm for 60 minutes. The total protein in the solution was 45 percent of the dry powder. The powder was found to retain its biological activities over a 14-month test period. Gel filtration was done on Sephadex G-50 and G-75 and ion-exchange chromatography was carried out on Cellex D. Behavior suggested the toxin was a basic protein with a molecular weight of 10,000–15,000. It was removed from DEAE-cellulose at basic pH and low strength.

Following poisoning of the crayfish *Orconectis virilis* with the acetone powder, the crustacean exhibited a paralysis characterized by an initial spastic phase, followed by a flaccid phase. Impaired reflexes were a constant finding. The righting reflex was used as a criterion for determining the degree of paralysis. Specimens which exhibited paralysis died within 48 hours. The immobilization dose (ID_{50}) for the acetone powder in crayfish was about 1 µg/kg body weight and the yield from a 70 g anemone with 23 g of tentacles was 1 g of the acetone powder, or a sufficient amount to paralyze approximately 2100 kg crayfish. It was found that the acetone powder solutions gave results which were equal to freshly prepared tentacle homogenates. Homogenates of non-tentacular tissue did not show significant paralytic activity.

In aqueous solution the acetone powder solution lost 100 percent of its activity on heating to 100°C for 30 minutes. Pronase destroyed the activity; trypsin destroyed 80 percent of the activity after 15 hours incubation at 23°C. The active component was nondialyzable. Gel filtration of the acetone powder solution on G-50 showed that the activity was contained in a single symmetrical peak at 2:12 times the void volume. The average peak volume was 2:16 times the void volume and recovery was 85–90 percent. Similar runs with crude tentacle homogenates on G-75 gave almost identical results. Toxic activity did not coincide with any peak in the total protein. The active portion was 75 times more potent than the acetone powder solution. It was found that the contaminating proteins were lighter than the activity fraction and that they could be separated on Sephadex G-50. Chromatography on G-50 followed by chromatography on G-75 increased the purification of the acetone powder approximately 150-fold.

On Cellex D, the recovery of activity was dependent on the conditions of attachment. When G-50 purified acetone powder was loaded at $\gamma/2$ (0.0225), the toxic activity came off as a sharp band between ionic strengths of 0.062 and 0.088. The increase in specific activity was 21-fold greater than the G-50 purified acetone powder. Stepwise changes in ionic strength were found to be more reliable: loading acetone powder or G-50 purified acetone powder at an ionic strength of 0.06 and removing it at 0.08 gave yields of 75 percent and purification of 25- to 27-fold. Purification using both gel filtration and ion exchange gave a product with an immobilization dose (ID_{50}) of approximately 1 µg protein/kg crayfish. This purified toxin was non-dialyzable, destroyed by pronase, and had a half-life in phosphate buffer of 24 hours at 4°C. It was markedly inactivated by shaking and foaming, and lost all activity within 10 minutes at 100°C. The behavior of the toxin, estimated from the rate of migration on columns indicated that it was a basic protein with a molecular weight of 10,000–15,000.

Further studies on a crayfish preparation showed that the toxin had a direct effect on the crustacean nerve but not on the muscle membrane. No evidence for a truly synaptic effect was observed. The toxin caused the motor axons to tire repetitively in high frequency bursts or low frequency trains. The spastic phase of paralysis was followed by a flaccid phase, attributed in part to a neural conduction block. Using the lobster giant axon and crayfish slow-adapting preparations, it was found that the toxin transformed action potentials into prolonged plateau potentials of up to several seconds duration, and that the eventual conduction block was not due solely to depolarization. It was suggested that the plateaus were caused, at least in part, by a prolonged membrane permeability following the initial excitation.

Cariello *et al.* (1989) have shown that the neurotoxic polypeptide from the anemone *Calliactis parasitica* has an amino acid sequence quite different from that of other known anemone toxins. The polypeptide chain has 46 amino acid residues, a molecular mass of 4886 Da and a pI of 5.4. No cystine residues are present. Like other anemone toxins, it interacts in crustacean nerve-muscle preparations with axonal membranes, yielding a massive release of neurotransmitter substance that results in a strong muscle contraction.

The sea whip *Lophogorgia rigida* and other gorgonian species have been shown to have an interesting toxin, lophotoxin. Extracts of *L. rigida* were chromatographed over magnesium silicate adsorbent and eluted with ethyl acetate in isooctane. Further purification was carried out by high performance liquid chromatography. Chemical analysis was consistent with $C_{22}H_{24}O_8$. In addition, a small amount of cembrenolide, pukalide (2) was obtained (Fenical *et al.*, 1981). This had previously been found by Scheuer (1975) in the soft coral *Sinularia abrupta*. The toxin had a subcutaneous LD_{50} of 8.9 mg/kg body weight in mice and blocked the indirectly elicited contractions in a mammalian nerve-muscle preparation while not affecting the directly elicited contractions. In the frog rectus abdominis preparation, ACh-induced contractions were blocked but KCl-induced ones were unaffected. It was concluded that lophotoxin produces an irreversible postsynaptic block, although the possibility of a presynaptic function could not be excluded (Culver and Jacobs, 1981).

In the sea anemones *Aiptasia pallida* and *Metridium senile* and in *Physalia*, Brand *et al.* (1993) isolated the low molecular weight collagens linked by disulfide bonds on SDA-PAGE and showed that there was a common major collagen species (40 kDa) for each nematocyst type.

Turlapaty *et al.* (1973) described a central nervous system stimulant in the form of a basic polypeptide isolated from the homogenized tissues of the anemone *Stichodactyla (Stoichactis) kenti.* The stimulation was described as "fighting episodes", and the authors suggested that the action was due to a central adrenergic mechanism. Devlin (1974) studied a partially purified toxin from *S. helianthus* which had a mouse intraperitoneal LD_{50} of 0.25 mg/kg and possessed hemolytic properties inhibited by sphingomyelin (Bernheimer and Avigad, 1976). Michaels (1979) and Shin *et al.* (1979) have shown that the cytotoxic poison from this anemone acts upon black lipid membranes and liposomes by channel formation and detergent action. This mechanism is also suspected for the hemolytic activity of some of the other sea anemone "cytolytic" toxins. However, there appears to be a considerable number of different cytolytic toxins in the sea anemones, and the mechanisms by which they damage membranes may be quite different. Hessinger and Lenhoff (1976), for instance, demonstrated that the toxin of *Aiptasia pallida* caused lysis through the action of phospholipase A on membrane phospholipids, while other workers suspected ionic or non-enzymatic roles.

Mebs and Gebauer (1980), in a well-executed group of experiments on a *Stichodactyla* species, separated biologically active polypeptides by gel filtration and ion-exchange chromatography. Following gel filtration on Sephadex G-75, two toxins were obtained, one having hemolytic activity and the other causing death in mice and crabs and inhibiting the proteolytic activity of trypsin and chymotrypsin. The hemolysin was further purified by chromatography and separated from the proteinase-inhibiting activity, which was resolved into three inhibitors. The molecular weight of the hemolysin was approximately 10,000; it had no phospholipase A activity, nor was lecithin required for hemolysis. Complete hemolysis of 0.6 percent human erythrocyte suspension was achieved at a concentration of 0.86 µg/ml. The hemolysin killed 1 cm fish (*Poecilia reticulata*) within one hour at a concentration of 4–5 µg/ml tank water. It was not deleterious to mice at 10 mg/kg body weight.

The lethal activity in the mouse was eluted from columns with proteinase inhibitors and had a suggested molecular weight of 6000. Although Mebs and Gebauer were unable to perform definitive lethality determinations in mice or crabs, they did find a subcutaneous minimum lethal dose of approximately 2.3 mg/kg for mice and an intramuscular minimum lethal dose of about 0.5 mg/kg for crabs. The amino acid composition of inhibitors 2 and 3, calculated on a basis of a molecular weight of 6000, gave a weight of 5800 for the main fraction. Inhibitor 2 had four half-cysteine residues and inhibitor 3 had six. The terminus was masked.

Kem *et al.* (1989) and Pennington *et al.* (1990) found that the tentacles of *Stichodactyla helianthus* contained a 48-amino acid residue "neurotoxin." The small protein was bound to the neuronal voltage-gated sodium channel, reducing channel inactivation and delaying the repolarization phase of the action potential. The secondary third-dimensional structure of the polypeptide, "neurotoxin I," the 48-residue toxin, consisted of a three-dimensional core of twisted four-stranded antiparallel B-sheet, encompassing residues 1–3, 19–24 and 40–47. These are joined by three loops, two of which were well defined. There was no evidence of an α-helical structure (Fogh *et al.*, 1990). According to Doyle *et al.* (1989), the cytotoxin of the anemone appears initially to bind reversibly to cell and bilayer membranes in a non-specific manner; sphingomyelin dominates in the bilayer and then provides optimal conditions for insertion into the membrane and subsequent assembly of a stable multimeric complex which functions as an ion channel.

The various polypeptide toxins of sea anemones that affect the sodium channel differ in their amino acid sequence, immunoreactivity and receptor binding sites, and have been divided into two separate groups (Schweitz *et al.*, 1985; Kem, 1988). The first type occur in the family Actiniidae, and includes the toxins from *Anemonia* and *Anthopleura*. Type 2 toxins occur in the family Stichodactylidae. They differ from the first type in lacking the N-terminal amino acid, possessing three consecutive acidic residues at positions 6–8, a single invariant tryptophan at position 30, and four consecutive basic amino acids at the C-terminus. Although the amino acid sequences of these two types of toxins are clearly homologous, only about 25 percent of the residues are identical. According to Hellberg and Kem (1990), in six anemone type I polypeptide toxins (46–49 residues) and in quantitative structure-activity relationships, 11 different amino acid positions may be of importance in determining physiochemical properties; positions 5, 21, 28, 34, 37 and 40, however were of primary importance in modeling the relative toxicities of the six polypeptides.

In 1967, Hashimoto and colleagues collected specimens of the file fish, *Alutera scripta*, from the Ryukyu Islands following a report that several pigs had died following their eating of the viscera of this fish. On examining the gut contents of the fish the investigators found polyps of the zoanthid *Palythoa tuberculosa* (Hashimoto *et al.*, 1969). About the same time, Scheuer and his group were investigating the toxin from *Palythoa toxica*, found at Hana off the island of Maui and called *limu-make-O-Hana* (death seaweed of Hana). Dr. H. Banner tells the story of how he and Dr. P. Helfrich first went in search of the organism on Maui in 1961 (where they found that the culprit was not an alga but a zoanthid). Much superstition was associated with the organism and "anyone who collected *limu-make-O-Hana* will meet with disaster," so believed the native islanders. When Banner and Helfrich returned to their laboratory on Coconut Island that evening they found that the laboratory building housing their facility had burned to the ground. Needless to say the islanders entertained their own etiology.

Palytoxin was first isolated from the zoanthid *Palythoa tuberculosa* by Moore and Scheuer, (1971). It was found that this organism was a source of the toxin in ciguateric fishes (Moore and Scheuer, 1971; Attaway and Ciereszko, 1974; Scheuer, 1978; Hashimoto, 1979). The toxin has also been found in two species of xanthid crabs (Raj *et al.*, 1983) and a red alga (Maeda *et al.*, 1985). The original purification method involved extraction from fresh, ground specimens in 70 percent ethanol, separation with DEAE Sephadex A-25, CM Sephadex C-25 and polyethylene gel. In 1983, Béress improved on the original technique, as did Hirata *et al.* (1979) and Uemura *et al.* (1985). Originally, the molecular weight was estimated at about 3300 (Moore and Scheuer,

1971) but revised by Macfarlane *et al.* (1980) to 0.2681 U
(12_c = 12.011 U). This gave a final molecular formula
$C_{129}H_{223}N_3O_{54}$. The ^{13}C NMR spectra were first described by
Moore *et al.* (1975) with a variant noted by Hirata *et al.*
(1979). The final structure was determined by Uemura *et al.*
(1981) and Moore and Bartoline (1981):

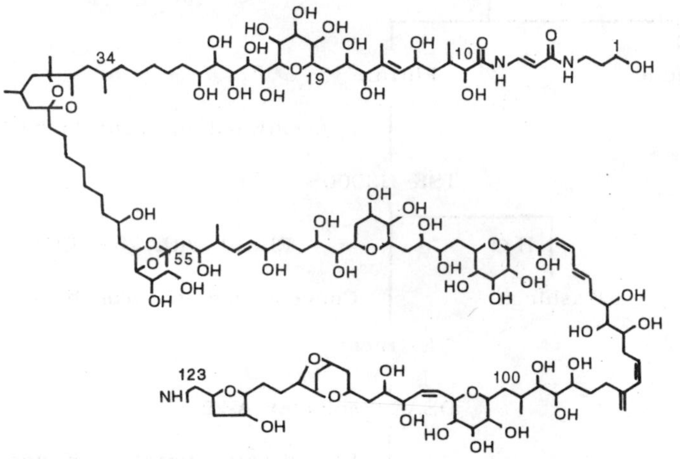

The two playtoxins from the Hawaiian *P. toxica* are related
C-55 hemiketals (Partial structures from Moore and Baroline,
1981). Their separation scheme is shown in Chart 2. Further subunits have been described by Uemura *et al.* (1985).

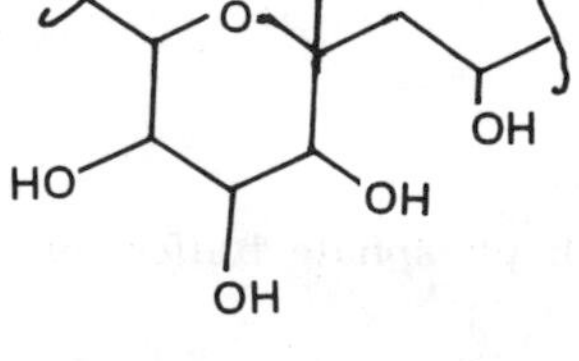

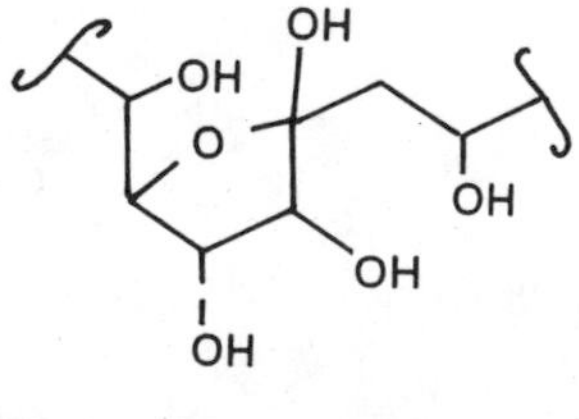

Some pharmacological properties of palytoxin have been
noted by Wiles *et al.* (1974) and Vick and Wiles (1975). The
i.v. LD_{50} in various mammals varied from 0.025–0.450
µg/kg body weight. There was a marked difference when the
toxin was administered intragastrically. Autopsy examination following lethal doses revealed that the toxin was
extremely hemorrhagic, primarily affecting the kidneys, but

Chart 2. Preparation of palytoxins (from Uemura *et al.*, 1985).

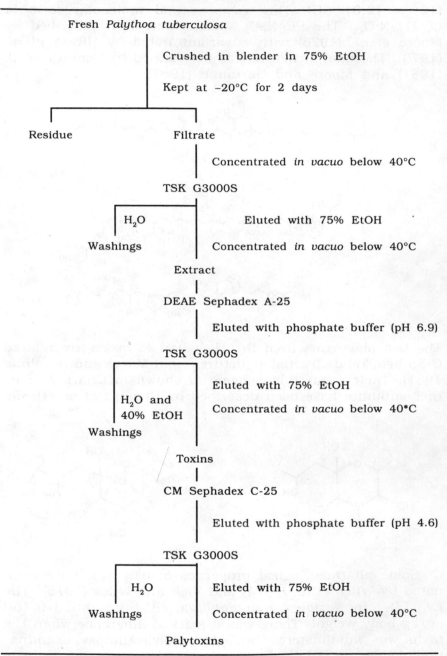

Fresh *Palythoa tuberculosa*

Crushed in blender in 75% EtOH

Kept at –20°C for 2 days

Residue   Filtrate

Concentrated *in vacuo* below 40°C

TSK G3000S

H₂O   Eluted with 75% EtOH

Washings   Concentrated *in vacuo* below 40°C

Extract

DEAE Sephadex A-25

Eluted with phosphate buffer (pH 6.9)

TSK G3000S

                       Eluted with 75% EtOH
H₂O and
40% EtOH   Concentrated *in vacuo* below 40*C

Washings

Toxins

CM Sephadex C-25

Eluted with phosphate buffer (pH 4.6)

TSK G3000S

H₂O   Eluted with 75% EtOH

Washings   Concentrated *in vacuo* below 40°C

Palytoxins

also the cardiovascular, gastrointestinal and respiratory systems. They concluded that death was due to renal failure, generalized hemorrhagic diathesis and necrotizing vasculitis, all leading to cardiac failure, shock and eventual death.

D. Clinical Problem

1. *Epidemiology*

It has been known at least since the days of Pliny that the stings of certain cnidarians can penetrate the human skin. While most nematocysts are capable of piercing only the thin membranes of the mouth or conjunctiva, some possess sufficient force to pierce the skin of the inner sides of the arms, legs and the more tender parts of the body. Still others can penetrate the thicker skin of the hands, arms and feet, and swallowing the tentacles or even the umbrella can cause epigastric pain and discomfort (Russell, 1965a,b; Halstead, 1965–70). Although stingings are usually associated with contact with the tentacles, Figure 25 demonstrates that effective nematocyst discharge can occur from the bell. In 1965, approximately 70 of the 9000 species of cnidarians were noted to have been involved in injuries to man. More recent records would indicate that about 95 species have

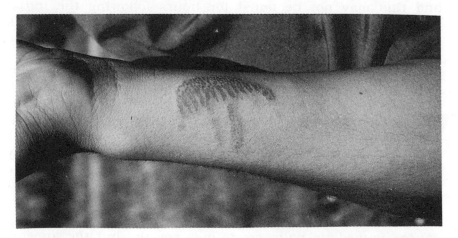

Figure 25. Lesion produced by jellyfish showing sting areas produced by bell as well as tentacles. (Marétić/Russell)

been implicated in such injuries (Russell 1965a,b; Halstead, 1965–70; Hashimoto, 1979; Maretić and Russell, 1983; Russell, 1984).

The number of envenomations by cnidarians is not known. Figures gathered between 1980 and 1983 would indicate that at least one and a half million persons are stung each year by these animals (Russell, 1984). Incomplete records gathered in India during 1990–1991 would seem to indicate that 7,000–10,000 persons are stung each year in Indian waters. Although the number of persons stung by cnidarians and subsequently treated by a physician is not known, estimates from the lifeguard groups along the southern California coast note that about one of 10 victims between 1955 and 1960 were referred to a physician or medical center. In Australia, however, the number of referrals would be much higher because of the more severe envenomations in that country.

2. Symptoms and Signs

The cutaneous lesions, as well as other clinical manifestations produced by the various cnidarians vary considerably, depending on the species inovolved and the number of discharged nematocysts, among other factors. Contrary to common belief, the stings of some cnidarians produce little or no immediate pain. Often, itching is the first complaint and this may not be noted for hours following the initial contact. In the authors' experience, stings by hydroids usually produce little if any pain, although there may be subsequent localized discomfort. In most cases, the lesions produced by hydroids are minimal. The sting of the fire, or stinging corals, Millepora spp. is usually associated with some localized, pricking-like pain, generally of short duration and with some subsequent pruritus and minimal swelling. There may be small reddened, somewhat papular eruptions, appearing 1–10 hours after contact which usually subside within 24–96 hours (Fig. 26). In severe cases, the papules may proceed to pustular lesions and subsequent desquamation.

Contact with the Portuguese man-of-war, Physalia sp., causes immediate pain, sometimes severe, and the appearance of small reddened, linear papular eruptions. At first

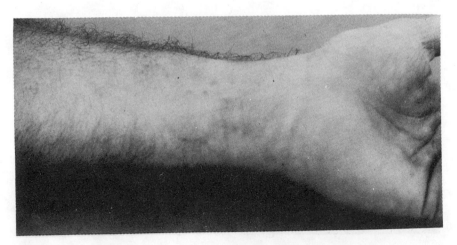

Figure 26. Lesions produced by a hydroid.

the papules are surrounded by an erythematous zone, but as their size increases the area takes on the appearance of an inflammatory reaction with small periodic, demarcated, hemorrhagic papules (Fig. 27). In some cases these papules are very close together, indicating multiple discharge of nematocysts as the tentacle passed over the injured part. The papules develop rapidly and often increase in size during the first few hours. The affected area becomes painful, and severe pruritis is not uncommon. Pain may spread to the larger muscle masses in the involved extremity or even to the whole body. Pain sometimes involves the regional lymph nodes. In some cases the papules proceed to vesiculation, pustulation and desquamation. In some cases, hyperpigmentation of the lesions is obvious for years following a stinging.

General systemic manifestations may also develop following *Physalia* envenomation. Weakness, nausea, anxiety, headache, spasms in the large muscle masses of the abdomen and back, vascular spasms, lachrymation and nasal discharge, increased perspiration, vertigo, hemolysis, difficulty and pain on respiration, described as being unable to "catch one's breath" cyanosis, renal failure and shock have all been reported (Russell 1965a,b: Russell 1966b; Dury *et al.*, 1980; Halstead, 1965–70; Burnett, 1989). Deaths from *Physalia physalis* stings are rare (Burnett and Gable, 1989; Stein *et al.*, 1989).

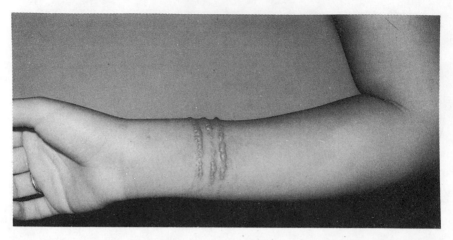

Figure 27. Linear papular lesions caused by *Physalia physalis*. (Russell/ Maretić)

Contact with most of the true jellyfishes gives rise, in the less severe cases, to manifestations similar to those noted for *Physalia*, with manifestations sometimes disappearing within 20 hours (Fig. 28). In the more severe cases there is immediate intense, burning pain, with contact areas appearing as swollen wheals, sometimes purplish, and often bearing hemorrhagic papules. These areas may proceed to vesiculation and necrosis. Localized edema is common and, in the more severe cases, muscle mass pain, difficulties in respiration, and severe spasms of the back and abdomen with vomiting are reported. Vertigo, mental confusion, changes in heart rate, and shock are sometimes seen (Barnes, 1960; Russell, 1965a,b; Sutherland, 1983).

The sea wasps, *Chironex fleckeri* and *Chiropsalmus quadrigatus* and certain other species are extremely dangerous Cubomedusae responsible for some deaths, particularly in the Indo-Pacific waters. The former, the more dangerous, may have a bell 15 cm across with tentacles, when their total length is added together, of over 90 m (Barnes, 1960). Systemic effects usually develop within 5–100 minutes following envenomation (Barnes notes an average onset of 20 minutes). Stings by these Cubomedusae cause a sharp prickling or burning sensation with the appearance of a wheal, which at first appears like a "rounded area of gooseflesh."

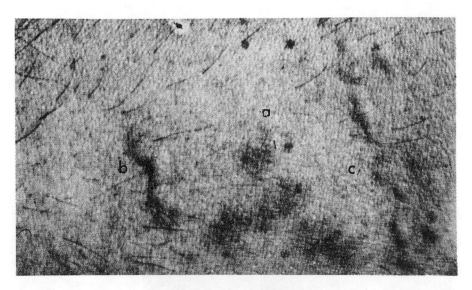

Figure 28. Experimental sting by *Carybdea rastoni* on human forearm. (a) Lesion at 24 hr, (b) Effect of 2 tentacles at 13 min, (c) Effect of single tentacle at 13 min. (Courtesy: R.V. Southcott)

An erythematous wheal soon develops and may become considerably larger than the area of contact. At first, it may show little pattern to suggest whether the stinging had been by tentacles or by the animals' umbrella. The wheals may either disappear, as when the stinging is minimal or, after an hour or so become enlarged as the nematocyst punctures become more apparent and appear as very small hemorrhagic vesicles surrounded by inflammation. A stinging pain develops and may persist for one to three hours. In linear lesions the nematocyst injuries may be no more than 5 mm wide but extend for 10 cm or more (Figs. 29, 30). Where the stingings have been from the umbrella the lesions appear as a cluster or an edematous wheal. Vesiculation and pustular formation may occur, and full thickness skin necrosis is not uncommon. Edema about the area may persist for ten or more days (Barnes, personal communication, 1967). Some distinguishing characteristics of stings by Australian cnidarians are shown in Table 11.

Cardiac disturbances, including failure, are not unknown following cnidarian stings. Martin and Audlay (1990) describe an interesting Irukandji sting (*Carukia barnesi*) in a 24-year-old female who experienced hypertension and atrial fibrillation. Neurological sequelae to jellyfish stings are also

Table 11. Characteristics of moderately severe Cnidarian stings off Australia.

Signs and symptoms	*Physalia*	Cubomedusae	*Cyanea*
Wheal			
Type	Single or replicated line	Multiple lines	Multiple lines
Width	Variable	3–7 mm	2–3 mm
Pattern	Irregular	Transverse bars	Zig-zag
Duration	1–4 hours	2–24 hours	30–60 min
Pain:			
Severity	Moderate	Marked	Less marked
Duration	30 min-2 hours	Many hours	10–30 min
Necrosis or vesication	Rare	Usual	Rare

Russell (1971a) and adapted from Barnes (1960).

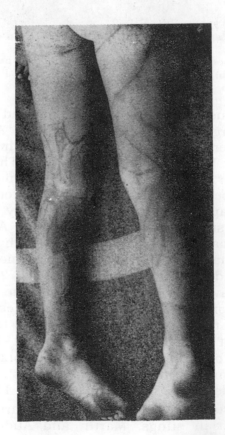

Figure 29. Linear lesions produced by tentacles of large *Chironex fleckeri*. showing hemorrhagic vesicles. Systemic manifestations may be severe. (Courtesy: J.H. Barnes)

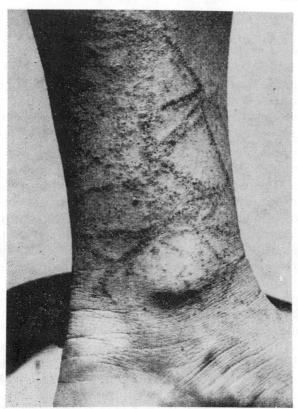

Figure 30. Lesions produced by *Chiropsalmus quadrigatus* at seven days. Injuries are less severe than from *Chironex fleckeri* but are nevertheless debilitating. (Courtesy: J.H. Barnes)

well known. In one case previously noted (Russell, 1966b) the patient developed ulnar nerve paresis and sensory loss but this was not observed at the time of the sting or reported to the author until a year later. The paresis was not evident in the first 48 hours following contact with *Physalia physalis* but developed on the third day and persisted for over a week. Peel and Kandler (1990) report a similar but more severe neurological deficit over many months following a stinging off Penang.

According to Menon (1930), the cnidarians, *Crambionella stahlmanni* and *Chrysaora quinquecirrha* are the most common stinging jellyfishes along the Madras Coast. Gravely (1941) reported that accidental contact with the tentacles of the above mentioned medusae will result in protracted pain. James *et al.* (1985) handled both species of live jellyfish for more than 30 minutes and while no pain or irritation were experienced during the handling, an accidental touch on the palm of his hand to the face caused a severe stinging pain of the eyelids. After one hour the face was washed with soap but the pain to both eyelids increased and persisted for several hours. This probably demonstrates that the nematocysts which stuck to the hand were transferred to the tender skin of the face and then to the eyelids while washing.

The sting of *Acromitus rabanchatu*, a common jellyfish of the Indian Ocean, particularly along the Indian coastline, is said to cause almost immediate pain, pruritus, redness and induration, sometimes paresthesia over the injured part, and occasionally headache, fever and shock. Some lesions become vesicular and papular, and occasionally break down becoming necrotic with subsequent sloughing and eventual desquamation. During a visit to India in 1989, one of us (FER) interviewed an 18-year-old male who had been stung by what appeared to have been *A. rabanchatu*. A thorough history and examination were done 24 hours after the accident. The youth had been bathing in shallow water at about 10:00 a.m. in the Bay of Bengal when he noted a "pricking sensation" over the calf of his left leg. He continued to bathe for about 10 minutes by which time the "prickling" had changed to pain and pruritus. On leaving the water, he immediately noted an increase in the itching and pain, and was told to rub wet sand on the area, which had now turned red and "puffy." The local manifestations were worsened by the rubbing of the wet sand.

At 30 minutes, the pain and itching were intense, and he was advised to stand in the cold surf, which offered some relief from the pain for a few minutes. About 45 minutes following the envenomation he was advised to see a local native practitioner. He ran to the hut of the practitioner and, on arrival, was hyperventilated and complained of nausea. He was given an unknown concoction and vomited 10 minutes later. A wet dressing of leaves and gauze was placed over the injured part, and he was given a drink which seemed to lessen the pain and made him "sleepy." When seen a day later, he had a reddened, slightly elevated area of approximately 4 × 9 cm over the left calf in which there were four well-defined lines of papular eruptions, a few of which had vesicated. His chief complaints were weakness, and pain and itching in the area of the stings. A balm consisting of hydrocortisone, diphenhydramine and tetracaine was applied to the lesion providing considerable relief. He was told to apply the balm every four hours and take 10 grains aspirin every six hours. Some of the vesicated eruptions subsequently ruptured and scabs formed, but otherwise recovery was uneventful. There was no report of headache, shock or other manifestations. The vomiting may have been due to the native drink.

Stingings by the anemones are usually of lesser consequences than those inflicted by the jellyfishes, and rarely are they as painful or disabling. The lesion area takes on a reddened and slightly raised appearance, bearing irregularly scattered pin-head size vesicles or hemorrhagic blebs (Maretić and Russell, 1983). They often appear flower-like (Fig. 31). The area becomes painful, particularly to touch or heat. In stings by *Anemonia sulcata* seen by one of us (FER) there has been some diffuse edema around the injured site. Residual hyperpigmentation is unusual following anemone stings.

Figure 31. Lesions caused by *Anemonia sulcata* seen at seven days. Note flower-like appearance. (Courtesy: Z. Maretić)

Stings by the stony corals (*Acropora*) give rise to some minor pain often followed by itching and the development of small diffuse wheals which may progress to vesiculation but rarely necrosis. "Sponge fisherman's disease" is thought to be due to the actinians *Sarortia elegans* or *Acropora rosea.* Troublesome are small spicules of coral which sometimes break off and become embedded in the skin, occasionally giving rise to infection.

On some Pacific islands, as well as in some other areas of the world, sea anemones are eaten following cooking, but some are apparently poisonous whether uncooked or cooked. *Rhodactis howesi* and *Physobrachia douglasi* are poisonous when eaten raw but said to be safe when cooked. *Heteroctis paumotensis* and related species are said to be poisonous whether raw or cooked. Intoxication is typified by nausea, vomiting, abdominal pain and hypoactive reflexes. In severe cases, marked weakness, malaise, cyanosis, stupor and death have occurred (L.C. Cutress, personal communication, 1958; Martin, 1960; Farber and Leske, 1963; Hashimoto, 1979).

3. *Treatment*

The number of remedies that has been suggested for the immediate first aid and subsequent treatment of coelenterate stings is impressive, ranging from alcohol and ammonia through Coca Cola, ice, fig juice and meat tenderizer to urine and vinegar, with several dozen other "treatments" between. It appears that some advice is animal species specific, other treatments are directed to immobilizing the nematocysts still in the skin, while still others are directed to reducing pain or itching. Obviously, still others are time-dependent or dependent upon what is at hand. Indeed, it seems best for the first attender or physician to seek advice from the expert in the area. Until there is a good deal more consistency in our knowledge about all of the parameters relating to cnidarian stings, his recommendations should be followed.

Unfortunately, most cnidarian stings occur in areas where medical advice is not easy to come by, and reliance is too often placed on local mythology or on persons not well informed. Such measures as rubbing the affected area with sand, washing with fresh water, immersing the part in hot water or wrapping the extremity in seaweed are still employed in some areas of the world, and are to be avoided. Besides being worthless, or contraindicated, these and other useless methods delay the institution of effective measures.

There is no question as to the importance of first aid and care. In well-controlled laboratory experiments in humans by Maretić and Russell (1975), they found that some manifestations (pain, itching or paresthesias) were present within two minutes of the nematocyst stings of *Pelagia noctiluca.* Dermatologic lesions were evident within 10 minutes, some within five minutes. One of us (FER) has experienced tingling within three minutes of handling the tentacles of *Gonionemus vertens*, this 15 minutes after the medusa was dead. There are other examples, particularly those cited in the works of Burnett and Williamson which indicate that poisoning can occur within minutes, so that the immediate care of the envenomation, particularly in cases of the box jelly (*Chironex fleckeri*), Irukandji (*Carukia barnesi*) and the Portuguese man-of-war (*Physalia* sp.) is essential.

It is not the purpose here to explore the many treatments for coelenterate stings. The interested reader is referred to the works of Southcott (1975), Russell (1965b), Halstead (1965–70), Fisher (1978), Hashimoto (1979), Pearn (1981), Sutherland (1983), Russell (1984), and Halstead (1988) for a more thorough review. The following measures may be appropriate for many cases but, again, there are specific therapies for the stings by some species and, as previously noted, the treating person should get advice on these from the expert in the area. The following measures, or their variations can be suggested for cnidarian stings in Indian waters.

1) Pour ocean water repeatedly over the injured parts. Do not use fresh water.

2) If ice is available, crush and place in a cloth over the stung parts.

3) Immerse part in vinegar for 30-60 minutes.

4) Apply some powder: flour, baking powder or even shaving soap to the affected parts and scrape the material off with a sharp knife or instrument. Do not use a razor.

5) Wash the areas again with salt water. Do not rub.

6) Wash, dry and apply a hydrocortisone (0.5%), diphenhydramine (2%) and tetracaine (1%) ointment in water base every four hours (the product used in the U.S. is called "Itch Balm Plus").

Systemic manifestations should be treated empirically and directed at the maintenance of ventilation, circulating blood volume and cardiac performance (Russell, 1976).

There is no question, as reported by C.E. Lane, F.E. Russell and B.W. Halstead at the First International Symposium on Animal Toxins in 1966, that salt water should be used in washing off any tentacles or unfired nematocysts from the skin. Fresh water is likely to cause the discharge of further nematocysts. The use of ice in the treatment of coelenterate stings probably enjoys a long history, although it is not mentioned in the exhaustive compendium by Halstead (1965–70). In 1958, at the Southern California Lifeguard Association meeting in Long Beach, California, Dr. Jerry Hughes noted that it was often the habit of the lifeguards to place ice over the stung parts when, in the case of the more severe envenomations, the victim was sent to a physician. The use of ice for the relief of pain has been known since antiquity. Cold, of course, is also a good vasoconstrictor, so its use in cnidarian stings has a basis in logic. Unfortunately, it has only been during recent years that the use of ice has found favor with physicians and those lifeguards who are generally the first attenders. This has been due, in part, to the fine report of Exton *et al.* (1989). As these authors have pointed out, however, its value is minimal in the presence of severe pain where codeine phosphate or other opiates may be required. There is no evidence that the ice in itself, affects the development of the lesions, although because of its vasoconstricting properties it may have some desirable action.

During a U.S. Office of Naval Research filming by P.R. Saunders and F.E. Russell in 1960–62, it was found that unfired nematocysts would adhere to baking power placed over the skin of affected areas. At the above noted Symposium, C.E. Lane suggested the use of aerosol shaving soap in the place of powder and this appeared to be used successfully by lifeguards in Florida. Although there is no uniformity of agreement on the application of vinegar, ammonia, alcohol or meat tenderizers, the authors prefer vinegar, which is also more commonly available for first aid. Alcohol, in our experience, appears to increase the pain, at least when applied locally. It has been said that by merely changing the pH of the skin, the interpretation of pain may be lessened. None of these substances, however, in our experience alters the development of the lesions once they have begun to appear.

The number of salves, balms, lotions, creams, oils ointment and sprays used for cnidarians stings is impressive. It appears that almost all of them have been recommended solely on the basis of their property to reduce pain, which is only one of the important manifestations of envenomation. None, except those containing corticosteroids, and perhaps diphenhydramine, appear to have any direct effect on the development of the lesions. In cases when there is extensive tissue breakdown, secondary infection may occur and an antimicrobial may be indicated. Hyperpigmentation can be treated with hydroquinone (Kokelj and Burnett, 1990) and keloid formation may respond to injectable steroids.

5

MINOR PHYLA

In this chapter we have included some of the minor invertebrates that might be noted as venomous, or possibly venomous, but about which little appears to be known concerning the toxicity of their poisons. Also, the clinical picture of "poisoning" by these animals is often vague, and it is sometimes difficult to decide whether or not the clinical manifestations are due to trauma, an allergic reaction or to a toxic secretion. Zoologically, we have tried to conform to a classification familiar to most biologists, particularly the non-specialist, and we apologize for taking the liberty of simplifying certain data. Further, we have introduced the various phyla alphabetically, for convenience. A modified listing of toxic species is shown in Table 12.

A. Annelida

The bristleworms, bloodworms and polychaete worms are elongated, segmented, bilaterally symmetrical worms with essentially similar segments that resemble each other in both their external appearance and internal structure. They also bear a well-developed head, a single preoral segment, the prostomium, a nervous system consisting of a pair of dorsal preoral ganglia (the brain), paired ventral nerve cords, a complete digestive system, and a closed circulatory system. They are typified by a nonchitinous cuticle secreted by the epidermis and supplied with chitinous bristles or setae. In some species there is a well-developed chitinous jaw. Most of the venomous marine species are found in the class

Table 12. Minor Phyla: Annelida, Bryozoa, Nemertini, (Platyhelminthes). Some species thought to be toxic.

Annelida, Class Polychaeta

    Family Amphinomidae
        Chloeia flava (Pallas)
        Chloeia viridis Schmarda
        Eurythoë complanata (Pallas)
        Hermodice carunculata (Pallas)

    Family Glyceridae
        Glycera dibranchiata Ehlers

    Family Lumbrineridae
        Lumbriconereis heteropoda Marenzeller

Bryozoa, Class Gymnolaemata

    Family Alcyonidiidae
        Alcyonidium gelatinosum Linneus
        Alcyonidium hirsutum (Fleming)

Nemertini, Class Enopla

    Family Amphiporidae
        Amphiporus angulatus
        Amphiporus lactifloreus (Johnson)

    Family Drepanophoridae
        Drepanophorus crassus (Quatrefages)

    Family Emplectonematidae
        Paranemertes peregrina Coe

    Family Lineidae (Class Anopla)
        Cerebratulus lacteus (Leidy)
        Lineus lacteus Montagu
        Lineus longissimus
        Lineus ruber

Platyhelminthes, Class Turbellaria

    Family Leptoplanidae
        Leptoplana tremellaris Oersted

    Family Planoceridae
        Stylochus neapolitanus Delle Chiaje

    Family Pseudoceridae
        Thysanozoon brocchi Grube

    Family Bdelloidea
        Bdelloura candida (Girard)

    Family Procerodidae
        Procerodes lobata

Polychaeta, which has approximately 5500 species. The annelids have a worldwide distribution and are, for the most part, benthic; most species are creeping or burrowing but some are sedentary, and a few are pelagic. They may reach a length of over 30 cm.

1. *Venom Apparatus*

Toxicity of the marine annelids is associated with their bristle-like setae or their biting jaws. Some species contain toxic substances within their bodies. With respect to the setae, these elongated chitinous bristles project from their parapodia. Each seta is secreted by a single cell at its base. In some species the setae can be extended or retracted so that with the worm at rest they appear short and sparse but when the animal is aroused the setae are extended and, as Halstead (1965–70) points outs, the animal appears to be a mass of bristles (Fig. 32). Eckert (1985) notes that the setae of *Eurythoe complanata* and *Pherecardia striata* are retractable; those of *Chloeia flava* are permanently everted and are hollow.

According to Halstead (1965–70), the hollow setae of *Eurythoe* and *Hermodice* are at times filled with fluid, but he was unable to show any glandular structures associated

Figure 32. A polychaete with thick tufts of parapodial setae (Courtesy: Smithsonian Institution).

with the setae, admitting, however, that the material he examined was poorly preserved. Eckert (1985) has examined fresh material and found a total absence of any gland cells directly associated with the setae. He concluded that the fireworms he examined were urticating rather than venomous.

Members of the genus *Glycera* have a proboscis equipped with four cuticular jaws, which can be everted rapidly to grasp or to tear food and pull the prey back into its mouth. The jaws have curved fangs, each connected by a duct to a venom gland. The gland lies within the pharynx (Fig. 33).

2.   Chemistry and Pharmacology

The marine annelid *Lumbriconereis heteropoda* was found by Nitta (1934) to contain a toxin in its epidermis. Nitta isolated a crystalline toxin which he called *nereistoxin* and subsequently reported on its pharmacological activities. In 1960, Hashimoto and his colleagues initiated a study on the toxin's chemistry, using the annelid *L. brevicirea*. Various papers described this poison as an amine with a 1,2-dithiolane ring. Its formula is $C_5H_{11}NS_2$, and proposed structure:

This structure was subsequently confirmed by Konishi (1970). On the basis of Nitta's observation that flies which came in contact with the worm lost their balance and often died, and the observations of others who noted the toxicity of extracts of the worm to arthropods, nereistoxin was studied as an insecticide. A large number of related compounds was synthesized, giving rise to the production of an important agricultural insecticide in 1972. Production of this insecticide reached an estimated 1,500 tons (Hashimoto, 1979).

The venom secreted from the venom gland of *Glycera convoluta* contains a group of macromolecular toxins. Michel and Keil (1975) characterized two proteolytic activities after separation of the secretion on Sephadex G-75 and G-200. A toxin with a suggested molecular weight of 110,000–

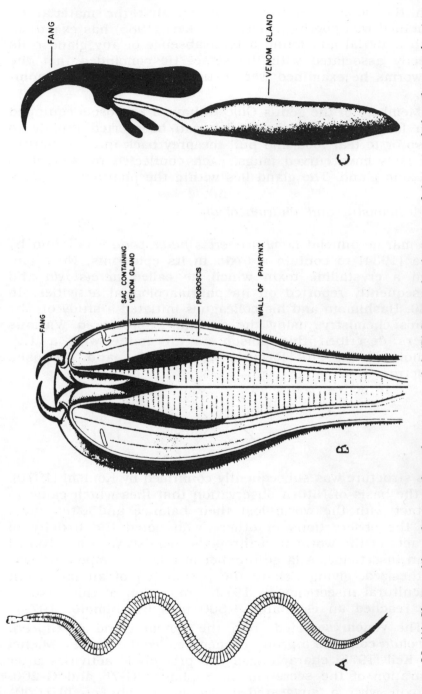

Figure 33. (A) *Glycera dibranchiata* with proboscis extended. (B) Longitudinal section of the extended proboscis showing two of the four fangs and related structures (C) Fang and venom gland. (From Halstead, 1965-70).

120,000 was identified after chromatography on agar-poly-acrylamide gel electrophoresis. It produced a paralytic effect on the heart of *Daphnia*. A group of proteinases, having molecular weights in the 60,000–70,000 range, degraded native insoluble collagen. Another proteinase of similar molecular weight acted on the synthetic substrate of trypsin, benzoly-L-arginine ethyl ester and was inhibited by a specific trypsin inhibitor.

Priapulus caudatus, a marine Ashelminthes of cylindrical shape and warty appearance, has been found to eject a brown viscous, foamable alkaline digestive fluid through its mouth. The material was found to have a protein concentration of 50 mg/ml fluid with a pH of 8.8–9.4. It had proteolytic activity with an optimum around 9.0; substrates specific for trypsin and chymotrypsin were hydrolyzed. No dipeptidase, lipase or amylase activity was observed. The toxin caused a sustained contraction of the isolated radular muscles of the whelk *Buccinum* sp. It was suggested that the strong proteolytic activity was associated with the animal's carnivorous nature.

3. *Clinical Problem*

The case for annelid setae being venomous appears to rest on several early clinical observations which are not too well documented and, in retrospect, might be questioned. Some early workers, however, have given good clinical descriptions which do implicate their poisonousness. Paradice (1924) records that large annelids (probably *Eurythoë complanata*) are covered with bristles and contact with them causes itching papules which last "the best part of a week". It was thought that "tiny siliceous" spicules which enter the skin and set up a severe irritation. Roughley (1940) reported that the setae can penetrate a leather glove and enter the finger tip. He wrote, "The tip of my finger remained numbed for six weeks." Penner (1970) found that the setae of the bristleworm *Hermodice carunculata* produced pain and transient numbness when they broke off in the skin. The pain or stinging sensation persisted for 13.5 hours and the numbness, which involved the entire extremity, persisted for about 30 minutes. He suggested that the setae contained a "neurotoxin," which was emptied into the wound when the bristles broke off. Other writers report such symptoms as "prickles, like a nettle sting," "burning irritation which lasts for hours," etc. Further descriptions of injuries have been provided by Pope (1947), LeMare (1952), Phillips and Brady (1953), Russell (1965a), Halstead (1965–70), Cleland and Southcott (1965), Hashimoto (1979), Halstead (1978), and Russell (1984).

Bites by some annelids, particularly large *Eunice aphroditois, Glycera dibranchiata, G. ovigera* and *Onusis teres* have been recorded as extremely painful and, in the case of *Glycera*, a venom can be involved. Cleland and Southcott (1965) have also noted a fisherman with a contact dermatitis who handled annelid worms. Eating these worms, of course, can be dangerous and cause illness.

Treatment of bristle stings is purely symptomatic. Systemic symptoms appear to be rare. Placing adhesive tape over the affected area to adhere to the setae, or even cellulose tape, has been suggested. Various ammonium and vinegar solutions have been advocated but do not appear to give much relief. Since the injury is of short duration and rarely disabling, little treatment is necessary.

B. Arthropoda

The arthropods (*Gr. arthros* = joint + *podos* = foot) make up the largest phylum in the animal kingdom. There are more than 1,000,000 species and some kinds are the most abundant animals on earth. The arthropods include the crabs, shrimps, barnacles and other crustaceans, insects, spiders, scorpions, ticks and their allies, centipedes and millipedes, among others. Their bodies are bilaterally symmetrical and segmented externally in varying degrees. They often show a divided head, thorax and abdomen. The appendages are jointed and external surfaces are covered by an organic exoskeleton containing chitin. They have a complete digestive tract with a terminal anus, a lacunar circulatory system, and they respire by means of gills, air ducts, book lungs or through their body surfaces. Some of the marine arthropods of the Indo-Pacific and Indian Ocean reported to be toxic are shown in Table 13.

Table 13. Some marine arthropods of the Indian Ocean and Indo-Pacific region reported to be toxic.

Class	Genus and Species
Merostomata (Horseshoe crabs)	
	Carcinoscorpinus rotundicauda (Latrille)
	Bay of Bengal to Phiippines
	Tachypleus gigas (Müller)
	Bay of Bengal to Vietnam and Torres Straits
	Tachypleus tridentatus (Leach)
	Indo-Pacific
Crustacea (Crustaceans)	
	Atergatis floridus (Linnaeus)
	Indo-Pacific
	Birgus latro (Linnaeus)
	Indo-Pacific
	Carpilius convexus (Forskål)
	Indo-Pacific
	Demania alcalai Garth
	Indo-Pacific
	Demania toxica Garth
	Philippine Islands
	Eriphia sebena Shaw and Noddler
	Indo-Pacific
	Lophozozymus pictor (Fabricius)
	Indo-Pacific
	Platypodia granulosa (Rüppell)
	Indo-Pacific
	Zosimus aeneus (Linnaeus)
	Indo-Pacific

1. *Chemistry and Pharmacology*

With the studies of Sommer (1932) and others, it became evident that the basic poison was paralytic shellfish poison (PSP), saxitoxin, *Gonyaulax* or dinoflagellate poison. Toxicity of the sand crab *Emerita analoga* was confirmed experimentally by Sommer in 1932. The toxin was found to be paralytic shellfish poison (PSP). Macht *et al.* (1941) found that at certain times of the year, muscle extracts from certain lobsters and shrimp were toxic to mice. Another report too frequently overlooked is that of Banner and Stephens (1966) who found an effective method for extracting toxin from the horseshoe crab *Carcinoscorpinus rotundicaudata* and *Tachypleus gigas* that was toxic to mice.

Hashimoto *et al.* (1967) screened various tissues of the crabs *Zosimus aeneus* and *Platypodia granulosa* following sporadic outbreaks of crab poisoning in the Ryukyu and Amami Islands, and found a poison they assumed to be ciguatoxin. They noted that "the difference in pharmacological and chemical nature between the toxin in ciguateric fishes and in our crabs may be too great to allow us to accept the prevalent belief of the inhabitants in the Ryukyu and Amami Islands, who attribute the ciguatoxin in fish to these crabs." In the following year, Konosu *et al.* (1968) identified the toxin as saxitoxin, and Mori *et al.* (1968) found that extracts of several crab species affected nerve membranes in mice and frogs.

Atergatis floridus has been found to be toxic (Inoue *et al.*, 1968). The toxicity of some *Demania* species was demonstrated by Alcala and Halstead (1970) and by Garth and Alcala (1977). The coconut crab *Birgus latro* was shown to be toxic to humans by Hashimoto *et al.* (1969) and by Bagnis (1970), while Fusetani *et al.* (1980) studied 10 percent acetic acid extracts of the hepatopancreas and intestines of the crab and demonstrated their toxicity in mice.

Hashimoto (1979) found that the distribution of crab toxin in the various parts of the animal differed remarkably (Table 14). If the lethal dose for saxitoxin to humans is used as a reference, he notes that a 0.5 g portion of the muscle of the chela of *Z. aeneus* may contain sufficient poison to kill a person. Teh and Gardiner (1974) found that the crab *Lophozymus pictor*, found off the coral reefs of Singapore, was toxic. An aqueous extract of the crab was subjected to a six-stage purification process, including separation on Sephadex G-50, and found to have an LD_{50} of 0.377 mg/kg in mice, with a rate-dose relationship that differed markedly from saxitoxin and tetrodotoxin. They suggested a molecular weight of 1,000–5,000 for their toxin, which gave reactions suggesting it contained free amino and phenol groups.

Llewellyn and Endean (1989a) found that aqueous extracts of the crab *Eriphia sabana* were toxic to mice, producing signs similar to PSP or tetrodotoxin. Electrophoretic and TLC studies showed that the toxins behaved like saxitoxin, gonyautoxin-1 and gonyautoxin-2.

Chia *et al.* (1991) found that in the coral reef crab *Lophozymus pictor*, extracts from the gut and hepatopancreas were the most toxic, while the muscle was less poisonous and the carapace either mildly toxic or nontoxic. Crabs kept in the laboratory lost almost all their toxicity within 24 days. Lau *et al.* (1991) examined the toxins from alcohol and aqueous solutions of a number of crabs from near Singapore and found those from *L. pictor* and *A. floridus* to be the most toxic. Almost all toxins were heat labile, except that from *A. pictor*. Lim *et al.* (1991) found that the toxin extracted in boiling water from *A. pictor* had an epileptogenic effect in rats when injected intracisternally.

2. Clinical Problem

The history of marine crab poisoning dates back to 1886 (Cleland, 1916). Cook Island natives were reported to eat the white-shelled crab "*Angatea*" during certain times of the year for the purpose of suicide. At other times, the crab was said to be non-poisonous. Rich (1882) describes a "crayfish" poisoning involving about 100 persons at Table Bay, South Africa, while numerous other early reports of crab poisoning have been noted. The interested reader is referred to Halstead (1965–70), Cleland and Southcott (1965) and Hashimoto (1979) for a more detailed accounting of marine crab poisoning. There is little doubt, however, that many marine

Table 14. Distribution of toxin in the body of some toxic crabs.

Body part	Toxicity (mouse units/g)	Body part	Toxicity (mouse units/g)
Zosimus aeneus		*Atergatis floridus*	
Appendages		Appendages	
Exoskeleton of chela	2000	Exoskeleton	150
Muscle of chela	6000	Muscle	400
Exoskeleton of walking legs	2000	Cephalothorax	
		Exoskeleton	65
Muscle of walking legs	3500	Muscle	45
		Viscera	25
Cephalothorax		Gills	<20
Exoskeleton	2000		
Muscle	40	*Platypodia granulosa*	
Viscera	1300	Appendages	
Endophragm	80	Chelae	400
Gills	25	Walking legs	500
		Cephalothorax	110

From Hashimoto, 1979. References in original table.

arthropod poisonings have been due to bacterial food poisoning, or that the ingestion of crabs or crustaceans has precipitated allergic reactions. Much of the earlier clinical literature on arthropod poisoning reflects these types of disorders and gave credence to the common belief that many crabs were poisonous.

Fatal poisonings however, following meals of crabs are well known. Among those arthropods implicated are *Demania toxica, D. reynaudi, Carpilius maculatus, Lophozozymus pictor* and *Zosimus aeneus* (Llewellyn and Endean, 1989b). Hashimoto (1979) has noted several additional crabs that have been implicated in poisoning (Fig. 34). Most cases of crab poisoning are relatively mild and the syndrome is easily differentiated from those caused by bacterial or allergic shellfish poisoning. The onset of symptoms usually occurs within three hours with nausea, followed by severe vomiting and diarrhea. Dizziness, drowsiness, weakness of the limbs, numbness of the lips and various other paresthesias, aphasia, muscle paralysis, a feeling of burning in the throat and stomach, and coma have all been reported. Manifestations are dependent upon the specific crab involved and the amount consumed. In 650 cases reported in the literature, there have been 43 deaths. Treatment is purely symptomatic.

C. Bryozoa (Ectoprocta, Polyzoa)

Bryozoans (*Gr. bryon* = moss + *zoon* = animal) are sessile tufted or branched cell colonies made up of individual zooides and inhabiting, for the most part, the seas and oceans where they are usually found in shallow waters. There are approximately 4,000 species of bryozoans, most of which are found attached to benthic animals or objects such as kelp, shells or rocks. Some species are mat-like, while others form encrustations. Bryozoans are often mistakenly included among the seaweeds and, while some resemble colonial hydroids and corals, their internal structure is far more complicated.

1. Chemistry and Pharmacology

In 1966, Turk *et al.* partially purified a toxic principle from *Alcyonidium gelatinosum* by column chromatography in

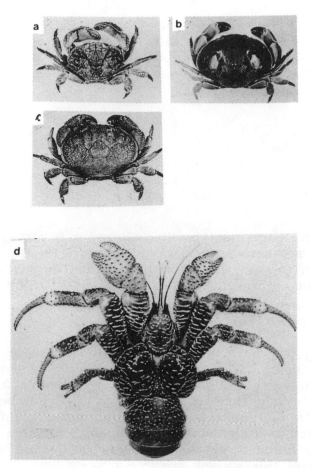

Figure 34. Toxic crabs. (a) *Zosimus aeneus*, (b) *Atergatis floridus*, (c) *Platypodia grannulosa*, (d) *Birgus latro*. (From Hashimoto, 1979)

Sephadex G-10. Four fractions were eluted. The first had an estimated molecular weight of greater than 700. This fraction and the third fraction proved to be active to sensitized guinea pigs. The third fraction, having a molecular weight of approximately 250–350, was more active than the first and appeared to contain acidic groups. It was suggested that the material might have an aromatic or heterocyclic structure. Subsequently, Carlé and his group at the University of Copenhagen carried out extensive chemical studies on the active toxic fraction of *A. gelatinosum*. The water soluble active portion was subjected to ethanol extraction, evaporation, ion-exchange chromatography, and nuclear magnetic resonance and mass spectrometry investigation. The data were consistent with the presence of a hapten (2-hydroxyethyl) dimethyl sulphoxonium:

$$H_3C \diagdown \atop H_3C \diagup S(\uparrow O) - CH_2CH_2OH$$

Due to its permanent positive charge it is strongly hydrophilic. A synthetic hapten was also produced. Approximately 5 ppm of the wet weight of the bryozoan is said to be the low molecular weight hapten (Carlé *et al.*, 1982).

2. Clinical Problem

Clinically, the most important bryozoan is the "curty weed" or "sea chervil," *Alcyonidium gelatinosum* or *A. hirsutum* of the family Alcyonidiidae, which is responsible for Dogger Bank itch. *A. gelatinosum* is found in great numbers off the Dogger Bank (Bonnevie, 1948), and more recently it has been noted as a problem in the Baie de Seine area (Audebert and Lamoureux, 1978) and off Denmark (Carlé *et al.*, 1982). Colonies of this species are free-growing and may reach a height of more than 50 cm.

There has been considerable confusion over the past five decades as to the causative organism of Dogger Bank itch. At various times the disorder has been attributed to a diatom, plankton, sponge, cnidarian or an alga. It now seems quite clear that the offending culprit is *A. gelatinosum* although many references to the disorder implicate *A. hirsutum. Fragilaria striatula* Lyngb. has also been implicated (Fraser and Lyell, 1963). The taxonomy and perhaps identification of the species are not firmly established (Thorpe and Ryland, 1979).

The bryozoan is generally taken in trawling nets and, while the fishermen are removing their edible catch, or while cleaning or repairing their nets, they come in direct contact with clusters of the organism. It is not universally accepted that the Dogger Bank itch is an allergic or sensitization phenomenon, although most evidence appears to indicate that this is so. Sensitization seems to vary with the extent and frequency of the exposure and, perhaps, with the individual. Continuous exposure from over six months to many years is the usual history given by fishermen who have the disorder. Once sensitized, the individual may develop allergic reaction in the presence of the bryozoan, whether it be on board ship, on the dock, or anywhere in close proximity to the organism. This would seem to indicate that the offending component of the bryozoan can be "released" from the organism and become airborne. On the other hand there is some clinical evidence that the disorder can occur from an initial contact with the animal. However, in such cases there is an indefinite history of previous exposure and often a history of many allergies, particularly to seafoods.

The incidence of Dogger Bank itch among North Sea fishermen is not known. However, of approximately 800 fishermen who had contact with the bryozoan and who were subsequently studied, about 10 percent had some history of the disorder. During the visit of one of us (FER) to Lowestoft, Copenhagen, and Ijmuiden during 1958, it was found that among the 50 fishermen who had handled the animal over 1 to 10 years, six had episodes of dermatitis on contact with the organism. It would be suspected that repeated handling of *A. gelatinosum* eventually leads to a high incidence of dermatitis. It is interesting to note that as far back as 1939 the Danish government passed an act, the Danish Workmen's Compensation Act, which included this disorder as an occupational hazard.

Clinical descriptions of Dogger Bank itch have been given by Bonnevie (1948), Seville (1957), Guldager (1959), Fraser and Lyell (1963), Russell (1965), Halstead (1965–70), Newhouse (1966), Audebert and Lamoureux (1978) and Carlé and Christophersen (1980). There is also a good review of the clinical problem in the discussion following the paper by Turk *et al.* (1966). Essentially, the disorder begins as reddening and slightly irritating or itching, usually on the flexor surfaces of the forearms and volar surfaces of the hands. On repeated or continued exposure, the dermatitis may spread to involve the arms, shoulders, chest and face. Severe pruritus and desquamation of the affected area may occur, and the involved parts may become quite painful to touch and to temperature changes. Angioneurotic edema and respiratory discomfort occur in the more serious cases, and joint involvement and wheezing have been reported (Russell, 1984). Identification of the active substance in Dogger Bank itch has depended upon patch testing in human subjects known to be sensitive to the bryozoan. Two patients, one suspected of having Dogger Bank itch and the other having had recurrences of the disorder over a 20 to 30-year period, displayed positive reactions with both the natural and synthetic haptens. In earlier studies (Turk *et al.*, 1966; Dubos *et al.*, 1977), sensitization of guinea pigs was used as a biological assay.

D. Nematoda

The nemerteans, or ribbon worms, are a small group of slender unsegmented worms distinguished from the Platyhelminthes by their rounded body, which usually tapers somewhat towards one or both ends, the presence of a prominent, evertible proboscis carried in a special chamber, the rhynchocoel, in some species, and a complete gut and circulatory system. These animals are carnivorous, feeding on annelids, crustaceans, molluscs and even fishes. Most species are marine. They vary in length from less than a millimeter to over a meter. The giant *Lineus longissimus* may reach 1 m. There are more than 15,000 living species.

1. *Venom Apparatus*

These worms may have a true venom apparatus (Kem, 1971). Envenomation by certain nemerteans occurs when the thread-like proboscis is extended explosively during the capture of prey or on handling. There is some question as to where the venom is produced and how it is transferred to the stylet when it is fired through the proboscis. It is known, however, that in feeding the worm jabs its prey with its proboscis stylet and the prey becomes paralyzed. To be effective this may need to be done several or many times (Fig. 35). In *Cerebratulus lacteus*, the toxin appears to be confined to peripheral tissues, principally the integument and the proboscis, both of which have dense populations of gland cells (Kem and Blumenthal, 1978a). For a more

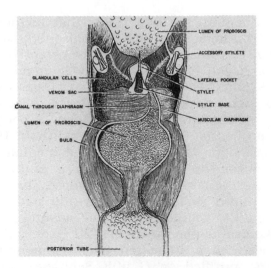

Figure 35. Semidiagrammatic drawing of stylet apparatus and associate structures in a nemetean worm. (From Halstead, 1965-70)

detailed discussion of this problem the reader is referred to Halstead's works and the fine review by Kem (1973).

2. *Chemistry and Pharmacology*

In 1936 and 1937, Bacq found two toxic substances in alcoholic tissue extracts of *Amphiporus lactifloreus, Drepanophorus crassus* and certain other species while studying the distribution of choline esters in various animals. He observed that when the extract he called amphiporine was injected into the lymph glands of frogs there was an increase in respirations, dilated pupils and extension of the limbs, with recovery within 30 minutes. The toxin did not seem to be localized in any part of the animal's body and was not released into the surrounding water by the worm. Bacq concluded that amphiporine was an alkaloid similar to nicotine. The second substance, nemertine, from *Lineus lacteus* and *L. marinus*, was said to be a "nerve stimulant" but no details of its properties were given. King (1939) attempted to isolate amphiporine from *A. lactifloreus* and carried out a number of extraction procedures but was unable to obtain a product suitable for analytical work. However, he noted that the action of amphiporine on certain pharmacological preparations was identical to that produced by nicotine.

It was not until the late 1960s that further studies were carried out on nemertean toxins. Kem (1969, 1971) and Kem *et al.* (1971) studied *Paranemertes peregrina, Cerebratulus lacteus, Lineus ruber, L. viridis* and *A. angulatus*. Toxin was obtained by mechanical or chemical (1 percent acetic acid) induction, which caused the release of copious mucus secretions. It appeared to be confined to peripheral tissues and the proboscis (Kem and Blumenthal, 1981). Quantal assays were done by measuring the median effective dose (PD_{50}) necessary to paralyze a 20 g crab. Paralysis was measured as the loss of righting ability. The secretions were then centrifuged, the supernatant adjusted to pH 5.0 and applied to a CM cellulose floc. The adsorbed toxin could be stabilized at a temperature of –20°C or on lyophilization. Extracts were then subjected to 1M ammonium acetate at pH 7.0 and purified on G 50F Sephadex and CM 52 cellulose column chromatography. Fractions were monitored at 280 and 230 nm, pooled and lyophilized. Assays were performed by measuring the PD_{50} necessary to paralyze a 20 g crayfish.

Kem (1976) found two main types of *Cerebratulus* toxins, one of which had almost all of the mouse lethality factor, toxin "A", and very little crayfish activity; and the other, toxin "B", which was very toxic to crayfish but not to mice. Further studies showed that toxin "A" contained three toxins, which Kem and Blumenthal (1978b) found to be very basic single chain polypeptides, having 94–98 amino acid residues and 3-4 disulfide bonds. On sequencing of one of these fractions, "A-III", it was found that the amino-terminal half possessed most of the charged residues, as well as four or six half-cystines, while the carboxyl-terminal half was substantially hydrophobic. This toxin rapidly lysed a variety of cells. It was specifically inhibited by sphingomyelin. Studies with human erythrocytes indicated that sensitivity was roughly proportional to membrane sphingomyelin: phosphatidylcholine ratios (Kem *et al.*, 1980). They suggested that since the three toxins had very similar amino acid compositions, they were probably "isotoxins" and, as preliminary amino acid sequencing findings indicated homology, and all three had potent cytolytic effects, including 50 percent lysis of human red cells in the 1–5 µg/ml concentration range, the toxins may be evolutionary variants of a common ancestral "A" toxin. Table 8 shows the amino acid composition of the toxins. It is refreshing to see this simple, applicable classification for these toxins, *Cerebratulus* A, A-II, A-III, etc., rather than what Hillard (1938) called the "scientific ego trip," the gimmick names so frequently used.

The manner in which *Cerebratulus* A toxins disrupt cell membranes has been the object of several studies and, again, it is refreshing to find Kem's statement that these toxins "should be designated as cytotoxins rather than hemolysins, as they lyse a variety of cells." We might go one step further and advise the word cytotoxic rather than cytotoxins, since the current authors are rather sure, from experiences with other toxins, that we will find that the *Cerebratulus* toxins have pharmacological properties other than their cytotoxic ones. Indeed, Kem and Blumenthal (1978b) note that sublytic concentrations of toxin A-III depolarize resting potentials and block action potentials in the squid giant axon and in the dog Purkinje cells. In the squid axon voltage clamp experiments, the sodium channel was completely blocked, even when depolarization of the resting potential was prevented by hyperpolarizing currents. Further study indicated that the depolarization was not due to the selective opening of sodium channels and that it may involve a non-specific increase in membrane permeability to several ions, including Na^+ and Ca^{++} (Kem *et al.*, 1980).

Posner and Kem (1978) pointed out the pharmacological similarities on rabbit atrial myocardial cells between A-III and "cardiotoxin" of the Formosan cobra; and Kem and Blumenthal (1978b) noted that sublytic concentrations of cytotoxin A-III inhibited voltage-dependent Na^+, K^+ and Ca^{++} channels in the squid axon. The reader might well ask if the fundamental mechanisms responsible for the deleterious effects of "cytotoxins", "cardiotoxins" and "neurotoxins" are not basically very much the same. Perhaps one should consider a receptor *mechanism* rather than a receptor *site*.

The four *Cerebratulus* B toxins were also shown to have large portions of basic amino acids (Table 8). It can be seen from the table that B-III and B-IV are very similar, both have four disulfide bonds and the common absence of three amino acids. They all appear to have molecular weights of approximately 6,000.

Following sequencing and further studies of B-IV, it was concluded that the Tyr-9 was directly related to the toxin-receptor interaction. B-IV prolongs repolarization of the action potential in the crayfish giant axon but this phenomenon is not seen in frog sciatic, garfish olfactory or squid axon preparations (Blumenthal and Kem, 1980). All four toxins produce similar effects in the crayfish, namely tremor, spastic convulsions followed by flaccid paralysis and death (Kem and Blumenthal, 1978b).

E. Platyhelminthes

The flatworms, Platyhelminthes, are characterized by having a flat and bilaterally symmetrical, unsegmented body, and having definite anterior and posterior ends. Some authorities note that they are the first animals to have a head and tail end. In the class Turbellaria, the epidermis is ciliated, while in Trematoda and Cestoda the surface is covered by a thick cuticle. Most species of Turbellaria are free-living, benthic and marine. Their epidermis is cellular or syncytial and has some cilia and rhabdoids. They have no respiratory or circulatory systems, anus or coelom, and are usually hermaphroditic. These flatworms vary in length from meters to less than a mm. There are approximately 10,000 species of Platyhelminthes, of which about 2,000 are Turbellaria.

1. *Venom Apparatus*

The turbellarians are thought to be the only group of flatworms that are toxic. There are certain glandular cells and rhabdoids in the animals' integument and both are secretory. They are thought to play a part in producing a toxin used in the animals' defensive armament. Although nematocysts have been found in and on turbellarians, they originate from the ingestion of hydroid tissues and pass through the mesenchyme and on to the epidermis.

2. *Chemistry and Pharmacology*

Arndt (1925) and Arndt and Manteufel (1925) studied crude extracts of the freshwater flatworms *Dendrocoelum lacteum, Polycelis nigra, P. cormuta, Planaria gonocephala, P. lugubris* and *Bdellocephala punctata.* They found that the extracts caused a reversible cardiac arrest in the frog heart. Arndt (1943) also studied crude saline extracts of some marine species, *Leptoplana tremellaris, Stylochus neapolitanus, Thysanozoon brocchi, Yungia aurantiaca, Bdelloura candida* and *Procerodes lobata* and found that they were toxic to guinea-pigs and that heating the extracts to 100°C for one minute destroyed the toxicity.

Thompson (1965) found a strongly acetic secretion, pH 1, in the epidermis of the worms *Cycloporus papillosus* and *Stylosatomum ellipse.* The secretion probably acts as a repellent to those animals which might try to feed upon the flatworm. The flatworms are of little clinical importance, although ingestion of the worms might cause poisoning. Recently, tetrodotoxin has been found in Turbellaria on Kyusya Island (Akakawa, *et al.,* 1991).

6

ECHINODERMATA

Echinoderms are exclusively marine animals, mostly benthic but found at all depths, and inhabit almost all water temperatures. They include the Asteroidea (sea stars), Ophiuroidea (brittle stars), Echinoidea (sea urchins and sand dollars), Holothuroidea (sea cucumbers), and the Crinoidea (sea lilies). In most cases they are characterized by radial or meridional symmetry, a calcareous endoskeleton made up of separate plates or ossicles which often bear external spines, a well-developed coelom, a water-vascular system and a nervous system, but no special excretory system. Of the 6,000 or so species, approximately 90 are known to be venomous or poisonous. The more important of these on a worldwide bases are noted in Table 15. Those from the Indian Ocean are identified with an asterisk.

The asteroids, starfishes, or more properly sea stars, have a central disc and five or more tapering arms or radii (Fig. 36). On the upper surface are many thorny spines of calcium carbonate in the form of calcite intermingled with organic materials. Between the spines are small, soft papulae that project from the body cavity and serve as gills and for excretion. Also around the spines are minute, pincerlike pedicellariae which serve to keep the surface free of debris and aid in the capture of food. The largest sea stars, such as *Pycnopodia*, have spreads of approximately 80 cm.

The sea cucumbers are soft-bodied animals covered by a leathery skin that contains only microscopic calcareous plates. The body is elongated in the oral-aboral axis. The mouth is surrounded by 10–30 tentacles, comparable with

Figure 36. The sea star *Astropecten polyacanthus*.

Table 15. Some echinoderms known to be venomous or poisonous. Those found in the Indian Ocean are noted with an asterisk(*).

Name	Distribution
ASTEROIDEA	
Acanthasteridae	
Acanthaster planci (Linnaeus)*	Widespread, Indo-Pacific, Polynesia to Red Sea, East Africa, North Australia, China, Japan, East Indies, Hawaii, Indian Ocean
Asteriidae	
Aphelasterias japonica (Bell)	Japan
Asterias amurensis Lütken	Japan
Asterias rubens Linnaeus	Iceland, British Isles to Senegal
Marthasterias glacialis (Linnaeus)	Eastern North Atlantic, Mediterranean to Cape Verde, Azores
Patiriella calcar (Lamark)	Australia, Tasmania, New South Wales
Pycnopodia helianthoides (Brandt)	California to Alaska
Asterinidae	
Asterina petinifera Müller and Troschel	Japan
Asterina burtoni Gray*	Indian Ocean
Astropectinidae	
Astropecten scoparius Valenciennes	Japan
Astropecten indicus Döderlein*	
*Astropecten polyacanthus** Müller and Troschel*	
Echinasteridae	
Echinaster sepositus Gray	English Channel to Cape Verde. Mediterranean
Echinaster purpureus (Gray)*	
Oreasteridae	
Pentaceraster affinis (Müller and Troschel)*	Indian Ocean
Goniasteridae	
Goniodiscaster scaber (Möbius)*	
Siraster tuberculatus H.L. Clark*	
Solasteridae	
Solaster papposus (Linnaeus)	Cicumpolar, European seas south to the English Channel
OPHIUROIDEA	
Ophiocomidae	
Ophiocomina nigra (Abildgaard)	Norway to the Azores, Mediterranean
Ophiocomina erinaceus Müller and Troschel*	

(contd.)

(Table 15. contd.)

Name	Distribution
Ophiocomina scolopendrina (Lemarck)*	

ECHINOIDEA

Arbaciidae

| *Arbacia lixula* (Linnaeus) | Mediterranean, West Africa, Azores, Brazil |

Diadematidae

Astropyga radiata (Leske)	
Centrostephanus rodgersi (A. Agassiz)	Australia, New Caledonia, New Zealand
Diadema antillarum Philippi	West Indies, South Atlantic to Brazil, East Atllantic
Diadema paucispinum (A. Agassiz)	Hawaii, South Pacific Islands
Diadema setosum (Leske)*	Indo-pacific, Polynesia to East Africa, Bay of Bengal, China, Japan, Red Sea, Persian Gulf, East Indies
Echinothrix calamaris (Pallas)*	Indo-Pacific, Polynesia to East Africa, Bay of Bengal, Australia, Japan, Red Sea, East Indies, Hawaii
Echinothrix diadema (Linnaeus)*	Indo-Pacific, Polynesia to East Africa, Australia, Japan, Red Sea, East Indies, Hawaii, Indian Ocean

Echinidae

Echinus acutus Lamark	European Seas, Iceland to Mediterranean
Paracentrotus lividus Lamarck	Atlantic coast of Europe to West Africa, Mediterranean, Azores
Psammechinus micro-tuberculatus (Blainville)	Mediterranean and Adriatic Seas, Azores

Echinometridae

| *Heterocentrotus mammillatus** (Linnaeus) | Indo-Pacific, Polynesia to East Africa, Hawaii, Australia, Red Sea, East Indies, China, Japan, Indian Ocean |
| *Echnometra mathaei* (de Blainville)* | Indian Ocean |

Echinothuridae

Araeosoma thetidis (Clark)	East coast of Australia, North Island of New Zealand
Asthenosoma varium Grube*	Indonesia, Indian Ocean, Suez (Red Sea), East Indies, China, Japan
Phormosoma bursaarium Agassiz*	Pacific, Indo-Pacific, Natal, (Hawaii to Arabian Sea)

(contd.)

(*Table 15. contd.*)

Name	Distribution
Spatangidae	
Spatangus purpureus Müller	Northern Europe to West Africa, Mediterranean Sea
Stomopneustidae	
Stomopneustes variolaris (Lamarck)*	Indian Ocean
Strongylocentrotidae	
Strongylocentrotus drobachiensis (Müller)	Circumpolar south to Scotland, Chesapeake Bay, Puget Sound
Strongylocentrotus purpuratus (Stimpson)	Vancouver to Lower California (North American West Coast)
Toxopneustidae	
Lytechinus variegatus (Lamarck)	West Indies, North Carolina, South Atlantic to Brazil
Sphaerechinus granularis (Lamarck)	Mediterranean, Channel Islands to Cape Verde, Azores
Toxopneustes elegans Döderlein	Japan, Ryukyu Islands
Toxopneustes pileolus (Lamarck)*	Indo-Pacific, Melanesia to East Africa, Japan, East Indies, Bay of Bengal
Tripneustes gratilla (Linnaeus)*	Indo-Pacific, Polynesia to East Africa, Red Sea, Australia, Japan, Bay of Bengal, East Indies, Hawaii
Tripneustes ventricosus (Lamarck)	West Indies south to Brazil, West Africa

HOLOTHUROIDEA

Name	Distribution
Chirodotidae	
Polycheira rufescens (Brandt)*	Japan, East Africa, East Indies, Bay of Bengal, Australia, Philippines, South Pacific
Stichopus chloronotus Brandt*	Indo-Pacific, Australia, East Africa, Bay of Bengal, East Indies, Japan, Hawaii
Stichopus japonicus Selenka	Japan, South East Asia, Indian Ocean
Stichopus variegatus Semper*	Polynesia to Indian Ocean
Stichopus naso Semper	Australia, Red Sea, East Africa, East Indies, Japan
Cucumariidae	
Cucumaria echinata (von Marenzeller)	Japan
Cucumaria fradatrit	
Cucumaria japonica Semper	South Pacific, East Indies, Japan
Neothyone gibbosa Deichmann	Gulf of California, Peru
Pentacta australis (Ludwig)	Tropical West Pacific, East Indies, Australia

(*contd.*)

(*Table 15. contd.*)

Name	Distribution
Pentacta armatus (von Marenzeller)*	
Thyone papuensis Theel*	Indian Ocean
Holothuriidae	
Actinopyga agassizi (Selenka)	West Indies
Actinopyga lecanora (Jaeger)*	Indian Ocean, Australia, Celebes, East Africa, Micronesia, Ryukyu Islands, East Indies
Actinopyga mauritina (Quoy and Gaimard)*	
Actinopyga miliaris (Quoy and Gaimard)*	East Africa, Indian Ocean
Bohadschia argus Jaeger*	Indo-Pacific, Polynesia, Ryukyu Islands, Bay of Bengal, East Indies, Australia
Bohadschia bivittata (Mitsukuri)	Philippines, Japan, China, South Pacific, Ryukyu Islands
Holothuria (*Halodeima*) *atra* Jaeger*	Indo-Pacific, Polynesia, Australia, East Africa, Red Sea, Bay of Bengal, East Indies, Japan, Hawaii
Holothuria axiologa H.L. Clark	Australia, Palau
Holothuria forskåli	
Holothuria (*Semperothuria*) *imitans** Ludwig	South Pacific, Indian Ocean
Holothuria (*Brandtothuria*) *impatiens** (Forskål)	Circumtropical
Holothuria (*Ludwigothuria*) *kefersteini* (Selenka)	Mexico to Peru and offshore islands
Holothuria (*Mertensiothuria*) *leucospilota* Brandt*	Australia, East Africa, Red Sea, Bay of Bengal, East Indies
Holothuria (*Selenkothuria*) *lubrica* Selenka	Gulf of California to Ecuador and Galapagos
Holothuria (*Selenkothuria*) *moebili* Ludwig*	Philippines, South Pacific, Hawaii, Japan, China, Indian Ocean
Holothuria monacaria (Lesson)	Fiji and Samoa, New Guinea, Timor, Queensland (Indo-Pacific)
Holothuria (*Microthele*) *nobilis** (Selenka)	Australia, East Africa, Red Sea, Ceylon, East Indies, China, Japan, Philippines, Hawaii, South Pacific, Indian Ocean
Holothuria poli Delle Chiaje	Mediterranean and adjacent Atlantic coasts
Holothuria (*Metriatyla*) *scabra** Jaeger	East Africa, Red Sea, Bay of Bengal, East Indies, Australia, Japan, South Pacific, Philippines, Indian Ocean

(*contd.*)

(*Table 15. contd.*)

Name	Distribution
Holothuria (*Theelothuria*) *spinifera* Theel*	Indian Ocean
Holothuria tubulosa Gmelin	Mediterranean, adjacent Atlantic coast
Holothuria vagabunda Selenka	Tropical western Pacific
Patinapta ooplax (von Marenzeller)	East Africa, Japan, East Indies, New Guinea
Molpadiidae	
Paracaudina australis (Semper)	Australia
Paracaudina chilensis (J. Müller)*	Japan, Australia, Bay of Bengal
Thelenota ananas (Jaeger)*	Indo-Pacific, Polynesia, Micronesia, Australia
Phyllophoridae	
Actinocucumis typicus Ludwig*	Indian Ocean, Pacific
Afrocucumis africana (Semper)*	Japan, East Africa, Bay of Bengal, East Indies, Australia
Euthyonidium sp. Deichmann	West Indies and American West Coast
Stichopodidae	
Parastichopus nigripunctatus (Augustin)	Japan
Stichopus chloronotus Brandt*	Indian Ocean
Stichopus variegatus Semper*	Indian Ocean
Synaptidae	
Euapta godeffroyi (Semper)*	Indian Ocean, Pacific
Euapta lappa (Müller)	Indian Ocean, Pacific

the oral tube feet of other echinoderms. The coelom is large and fluid filled. Movement of the muscles over the fluid-filled body enables the animal to extend or contract its body and to carry out wormlike movements. The largest sea cucumber, *Synapta maculata*, may measure 2 m in length when less than 5 cm in diameter. According to Nigrelli and Jakowska (1960) at least 30 species belonging to four of the five orders of Holothuroidea are toxic. Some toxic species, *Thelenota ananas, Stichopus variegatus, Holothuria atra* and *H. axiologa*, are esteemed as food in the Orient. The dried body wall of some sea cucumbers is often boiled and dried in the sun to produce "trepang" or "bêche-de-mer," used in soups.

A. Venom Apparatus

The fragile calcite spines of the sea stars, *Acanthaster* sp., are covered by a thin integument composed of an epidermis and a dermis. In the epidermis there are two kinds of cells, acidophilic and basophilic, among others. It is believed that a venom is produced in the acidophilic glandular cells, which are peripheral in their distribution in the epidermis. It appears that the toxin can be discharged into the water by the sea star or, as in the case of humans, directly onto the skin. Some asteroids are known to be poisonous when eaten.

The sea urchins, Echinoidea, have two main types of spines, primary and secondary (Fig. 37). The spines are composed of calcium carbonate in the form of calcite intermingled with organic material. In most echinoids the spines are solid and do not contain a venom gland. In Echinothuridae and Diadematidae, the primary spines are long, slender and hollow (Figs. 38a, b) and are easily broken off in a wound. They may attain a length over 30 cm. In the primary spines of *Phormosoma bursarium*, the secondary spines of *Araeosoma thetidis* and some others, the acute tips of some spines are encased within a venom sac. These have been described by Mortensen (1928–51), Hyman (1940), and Halstead (1965–70).

The principal venom apparatus in the sea urchin, heart urchin and sand dollar, however, is the pedicellaria. In essence, pedicellariae are modified spines with flexible heads.

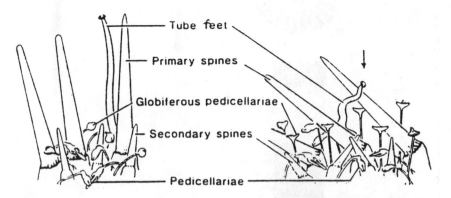

Figure 37. Diagrammatic sketch of relationships between primary and secondary spines, pedicellariae, and tube feet (from Russell, 1984).

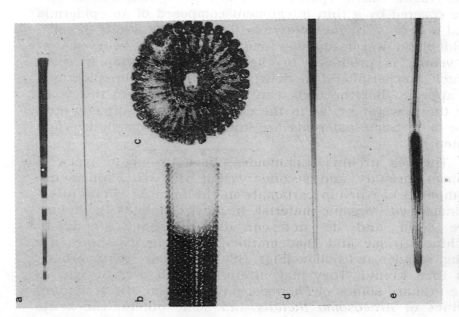

Figure 38b. *Echinothrix diadema*

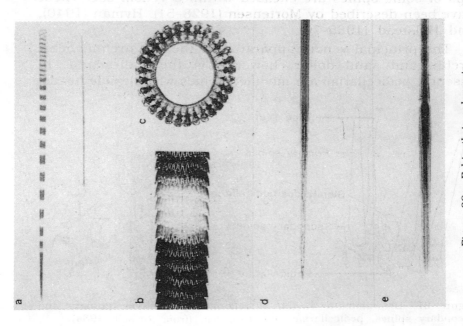

Figure 38a. *Echinothrix calamaris*

There are four primary kinds: ophiocephalous, dentate, globiferous and foliate. Some urchins possess all four kinds, although not all are found in every genus of sea urchin. The pedicellariae function in food getting, grooming and self-defense. The glandular, gemmiform, or globiferous type pedicellariae serve as venom organs. In most echinoids the so-called "head" of the pedicellaria is composed of three calcareous jaws or valves, each having a rounded, tooth-like fang. The jaws are usually invested in a globose, fleshy and somewhat muscular sac which possesses a single or double gland over each valve (Fig. 39).

O'Connell (1971) studied the fine structure of the globiferous pedicellariae of the purple sea urchin, *Strongylocentrotus purpuratus*. He found secretory cells lining the inside of the venom gland that were typically mucoid in appearance and possessed small volumes of basally displaced, vesiculated cytoplasm. They also had an extensive system of vacuoles which dominated the apical nine-tenths of each cell. The vacuoles contained ground substances of various densities. The venom-producing cells contained numerous cytomembrane complexes, indicating metabolic activity. The vacuolar products were tentatively identified as basophilic, fuchsin-positive spherules, apparently formed of protein, and dense, granular, alcianphilic compounds containing nonsulfated acid mucosubstances. O'Connell concluded that the gland cells were probably holocrine in function and released their vacuolar contents upon constriction of the muscles surrounding the gland sac.

In the sea cucumbers, Holothuroidea, some members possess special defense organs known as Cuvierian tubules, which arise from a common stem of the respiratory tree. When these animals are irritated they emit these organs

Figure 38a. (a) Primary spine and secondary spine with surface epithelium removed; (b) surface structure of primary spine showing spicules arranged in rows and columns; (c) cross-section of primary spine showing extensive lumen; (d) acute tip of secondary spine with epithelium removed and showing retrorse spinules; (e) acute tip of a secondary spine showing invested epithelial sac and retrorse spinules.

Figure 38b. (a) Primary spine and secondary spine with surface epithelium removed; (b) surface structure of primary spine showing spinules arranged in rows and columns; (c) cross-section of primary spine showing small lumen; (d) acute tip of a secondary spine with epithelium removed showing retrorse spinules; (e) acute tip of partially decalcified secondary spine showing epithelial sac (From Alender, 1967).

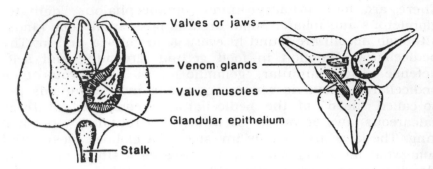

Figure 39. Head of pedicellaria showing three calcareous jaws, venom glands and associated structures (From Russell, 1984).

through the anus. The tubules become elongated by hydrostatic pressure so that once through the anus they become extremely sticky threads in which the attacking animal can become ensnared. As noted by Endean (1957), the process of elongation may split the outer layer of covering cells, thereby releasing a material which forms an amorphous mass having strong adhesive properties. These elongated threads may attain a length of almost a meter. They separate from their attachment and are left behind as the holothurian crawls away, although only parts of the tubules are emitted at a time.

In *Actinopyga agassizi*, as in some other sea cucumbers, however, the tubules do not become sticky, nor do they elongate, but they are eviscerated in a somewhat similar manner and discharge a toxin from certain highly developed structures filled with granules. The toxin is capable of killing fishes and other animals. Evisceration may be provoked by handling the animal, either in or out of the water, and by excessive changes in temperature, pH and oxygen balance. In other sea cucumbers, as in *Holothuria atra*, which do not possess Cuvierian tubules, the toxin may be discharged through the body wall. In some parts of the Pacific and Asia, the common black sea cucumber is cut up and thrown into tide pools, or squeezed into crevices so that the body fluids can deactivate fish. "Before long the water became noticeably turbid in appearance, and shortly thereafter fish began coming to the surface of the pool, exhibiting much the same type of behavior as in rotenone poisoning" (Frey, 1951).

B. Chemistry and Toxicology

Many echinoderms secrete a mucus or liquid from their integument that plays a role in their defensive armament. The chemical nature of these substances varies considerably as, obviously, does their pharmacological activity. For instance, the viscous discharge from the massive multicellular integumentary glands of the brittle star *Ophiocomma nigra* is characterized as a highly sulfated acid mucopolysaccharide, containing amino sugars, sulfate esters and other substances complexed to proteins (Fontaine, 1964). The pH of this discharge is approximately 1.0, which probably makes it very offensive to other marine animals which might seek to prey upon it. Among the other substances isolated from echinoderms are a quaternary ammonium base ($C_2H_2NO_2$, picoline acid methyl betaine), known as homarine (Hoppe-Seyler, 1933; Gasteiger *et al.*, 1960), several phosphagens (Phosphoarginine and phosphocreatine), and sterols (Bergmann and Domisky, 1960; Scheuer, 1973), saponins (Hashimoto *et al.*, 1960; Yasumoto, 1966; Ruggieri *et al.*, 1970), and other compounds (Hashimoto, 1979).

1. *Asteroidea*

Many of the sea stars (starfish) also release a bioactive substance from the epidermis of their tube feet (Feder, 1963) to which some molluscs, anemones and sea urchins respond (Feder, 1959; Montgomery, 1967; Hashimoto, 1979). The active substance is probably sensed by chemoreception, either in the water or through direct contact with the asteroid. This substance can not only repel certain marine invertebrates but it can be lethal to some fishes. Hashimoto and Yasumoto (1960) found a saponin-like substance, having a varying sugar composition but always bearing a sulfate, in the secretion released by the sea star. Rio and associates (1963) also found a saponin-like substance in the sunburst sea star *Pitasometra helianthoides*.

A toxic substance was isolated from dried meal of *Asterina pentinifera* that was suspected of being a saponin (Yasumoto *et al.*, 1966). Using specimens of fresh *Asterias amurenis*, they found that their partially purified product was toxic to killifish, lethal to fly maggots and earthworms, and caused vomiting in cats. On thin layer chromatography the toxin was resolved into six components (Yasumoto and Hashimoto, 1965). The major component was named asterosaponin A and the second major component was termed asterosaponin B. Their chemical properties are shown in Table 16. Using various chromatographies, Ikegami *et al.* (1972) obtained two bioactive fractions from sea stars that were similar to asterosaponins A and B.

Partially purified drippings from thawed *Asterias rubens* and *Marthasterias glacialis* caused marked avoidance reactions in *Buccinum* and several species of ophiuroids, and strong muscle contractions in *Buccinum undatum* radula muscle preparations. It was speculated that since the reactive substances are produced in the same external epithelium of the oral and aboral surfaces of sea stars, possibly all predatory sea stars of the order Forcipulata might exude similar substances causing avoidance reactions (Feder and Arvidsson, 1967).

Mackie *et al.* (1968) isolated and partially characterized a surface-active saponin from *Asterias rubens*. It had an approximate molecular weight of 514, and paper chromatography revealed that this major steroid was present with at least four other steroid components. The principal saponin, saponin 1, was found to provoke the greater part, if not all, of the activity of the extract toward the intact snail response. Its chemical characteristics seemed similar or

Table 16. Properties of asterosaponins A and B.

Property	Asterosaponin A	Asterosaponin B
Melting point	185–190°C	189–191°C
Specific rotation $(\alpha)_D^{20}$	+ 0.03°	– 2.66°
Hemolytic index	55600	41000
Molecular formula	$C_{50-52}H_{88-90}O_{25-26}SNa$	$C_{56}H_{101}O_{32}SNa$
U.V. absorption	244 nm	248 nm
I.R. absorption	1640cm^{-1}	1700, 1640 cm^{-1}
Sugar	D-Quinovose, D-Fucose (2:2)	D-Quinovose, D-Fucose, D-Xylose, D-Galactose (2:1:1:1)
Sulfate group	One mole	One mole
Aglycone	Asterogenin-I,II	Asterogenin-I,II

From Hashimoto, 1979.

identical to the asterosaponin described by Yasumoto and Hashimoto (1965). The active principle in *Marthasterias glacialis* also behaved in a similar manner chemically, and its foot-muscle bioassay effects were similar. A second saponin, saponin 2, had different characteristics. Mackie and Turner (1970) concluded that the saponins or saponin-like substances were surface active, eliciting either an immediate avoidance reaction or, as in the mollusc *Buccinum undatum*, a series of convulsions. The authors developed a sensitive, semi-quantitative assay using the withdrawal response of the intact mollusc's foot-muscle. Positive correlations between the intact preparation and the tube-foot reponse were demonstrated. A further bioassay, the isolated radular muscle of *Buccinum*, showed that in order to elicit a significant response the presence of a protein fraction was essential.

A "lethal" saponin fraction from *Asterias vulgari*, which contained four major components, was isolated by Owellen *et al.* (1973). It had cytolytic activity against mouse, rat and human erythrocytes and lymphocytes. Although an LD_{50} determination was not made, mice survived intraperitoneal doses of 50 mg/kg body weight but succumbed to 200 mg/ kg. Croft and Howden (1974) isolated and partially characterized a hemolytic steroidal saponin from the sea star *Patiriella calcar*, which they called calcarsaponin. Its aglycone, calcargenin, was partially characterized and had a molecular weight of 414. The sugars in calcarsaponin were galactose, fucose and quinovose, and the saponin contained a sulfate moiety. Fleming *et al.* (1976) found a mixture of four steroidal saponins in *Acanthaster planci*. The major sapogenins were 5α-pregn-9(11)-ene-3β,6γ-diol-20-one and 5α-cholesta-9(11), 20(22)-diene-3β,6γ-diol-23-one. Spectral evidence was obtained bearing on the structures of the remaining sapogenins. Hydrolysis of the saponins also yielded quinovose, fucose, arabinose, xylose, galactose, glucose and sulfate. The highly surface-acting properties of the asteroid saponins make them hemolytic, anticoagulant, cytolytic and toxic to fishes.

2. *Echinoidea*

One of the first observations that certain pedicellariae were venomous was probably that of Von Uexkull (1899), who demonstrated that when the pedicellariae of several urchin species were allowed to sting various animals signs of poisoning developed. Henri and Kayalof (1906) found that aqueous extracts of several urchins were lethal to certain gastropods, cephalopods, crustaceans, fishes, a lizard and rabbits but not to several echinoids or a frog. Levy (1925) demonstrated the hemolytic effect of several sea urchins' pedicellarial extracts, and Fujiwara (1935) observed that when the pedicellariae of *Toxopneustes pileolus* were allowed to sting the abdomen of a mouse the animal developed respiratory distress and exhibited a decrease in body temperature. Pérèz (1950) found that injection of thermostable extracts from the macerated pedicellariae of *Sphaerechinus granularis*, and certain other species, were lethal to isopods, crabs, octopods, sea stars, lizards and rabbits, although they had little effect on frogs. He thought that the toxic factor could be found in other tissues of the animal, although he suspected that more was concentrated in the pedicellariae. The presence of a dialyzable acetylcholine-like substance in the pedicellariae of *Lytechinus variegatus* was reported by Mendez (1963). The toxin was partially deactivated by heating and completely destroyed by one hour's exposure to NaOH.

The protein nature of pedicellarial toxin was described by Alender *et al.* (1965) and Feigen *et al.* (1966). They found that the active principle of *Tripneustes gratilla* was non-dialyzable, thermolabile, pH stable, almost completely soluble in distilled water, and precipitable with a relatively high potency in the presence of two-thirds saturated ammonium sulfate in a yield of 26 percent dry weight of the starting material. Sixty percent of the lethal activity appeared to be in one fraction that had a sedimentation coefficient of 2.6 Svedberg units at 20°C. The intravenous LD_{50} in mice, based on the quantity of precipitable protein nitrogen, was 1.59×10^{-2} mg/kg body weight for the crude material and 1.6×10^{-2} mg/kg for the protein.

This protein possessed hemolytic activity against human type A and B, rabbit, guinea pig, beef, sheep and fish erythrocytes. Intravenously, it produced a dose-related hypotension that was responsive to adrenalin. It had deleterious effect on the isolated heart and guinea pig ileum. In both these preparations the toxin caused the release of

histamine and serotonin. It seemed to have little effect on
isolated toad nerve (Alender and Russell, 1966). Using the
deep extension abdominal muscle of the crayfish, Parnas
and Russell (1967) showed that the toxin produced a rapid
block in the response of the indirectly stimulated muscle.
Even with low concentrations there was an irreversible block
in the muscle's response to intracellular stimulation. The
compound action potential of the crayfish limb nerve was
also blocked by the toxin but this potential reappeared on
washing. The toxin caused considerable damage to the muscle
fibers. These findings seemed to indicate that pedicellarial
toxin blocks the response from both nerve and muscle, and
that it is cytolytic. A similar conclusion was gleaned from
studies on the nerve-muscle preparation of the guinea pig.

The groups at Stanford (Feigen *et al.*, 1966) observed
that pedicellarial toxin from *Tripneustes gratilla* elicited
prolonged contractions of isolated guinea pig ileum. Chemical evidence was obtained for the release of histamine from
ileal, cardiac and pulmonary tissue, as well as from the
colonic and pulmonary tissues of the rat. The histamine
release was quantitatively dependent on the concentration
of the toxin acting upon the tissue. Subsequently these
workers showed that the toxin acted kinetically as an enzyme,
and that one of the substrates was electrophoretically pure
α_2-macroglobulin. They also demonstrated that the crude
toxin was kininolytic, with respect to the dialyzable reac-
tion-product, as well as to synthetic bradykinin.
Immunoelectrophoretic analysis revealed the existence of
two distinct antigenic determinants, and additional serologi-
cal tests showed that antibodies directed against the
pedicellarial toxin of *T. gratilla* cross-reacted with the test
proteins of *Strongylocentrotus purpuratus*. The active material obtained from the reaction between crude sea urchin
toxin and heated plasma was a mixture of pharmacologically
active peptides, one of which was bradykinin. The peptides
could be separated by gel filtration on Sephadex and resolved into seven distinct components by paper
chromatography, three of the slower moving peptides being
chromatographically and pharmacologically identical with
commercial bradykinin (Feigen *et al.*, 1968).

In 1974, Fleming and Howden obtained a partly purified
toxin from the pedicellariae of *T. gratilla* by extracting with

two volumes of cold 0.01 M TrisOHCl buffer, pH 8.0, homogenizing the product, centrifuging, decanting the supernatant and filtering through a Millipore filter. Ion-exchange chromatography on carboxymethyl cellulose and electrofocusing showed that the toxin was acidic in nature. The major components were eluted at 4.1, 5.1 and 5.4. The toxin, as established by intraperitoneal injection in mice, was at the 5.0–5.1 isoelectric point. When this fraction was chromatographed on Sephadex G-200, a molecular weight of 78,000±8,000 was obtained. This seems consistent with the sediment coefficient of 4.7 (67,000) presented by Feigen *et al.* (1968).

Kimura *et al.* (1975), perhaps unaware of the works of Feigen *et al.* (1974), also found that the pedicellarial venom of *Toxopneustes pileolus* was a mixture of several basic peptides having kinin-like activity. Following extraction and separating procedures, they found that their urchi-toxin F_2 exhibited absorption maxima at 260, 280 and 330 mμ. Ultracentrifugation yielded two peaks having sediment constants of 2.1 and 4.7, respectively. The approximately molecular weights were 30–40,000 and 70–80,000. Thin-layer chromatography indicated a single spot but with some trailing. The toxin was basic in character. It produced a decrease in frog heart activity and an irregular rhythm, caused transient vasoconstriction, increased motility of rabbit ileum, contraction of guinea pig ileum and rat uterus, and increased peripheral blood vessel permeability. The interacting activity was destroyed by chymotrypsin. On exposure of urchi-toxin F_2 to Sephadex G-25, three fractions (F_3, F_4, F_5) were obtained. Only F_3 had significant biological activity. F_3 had a lesser cardiac effect than F_2, while F_4 and F_5 had no effect. F_3 and F_5 produced contraction of the isolated guinea pig ileum. When these various properties were compared with those produced by bradykinin, the conclusion was that F_2 had 50 percent bradykinin activity, while F_3 had 75 percent, F_4 had 0 percent and F_5 had 30 percent.

In 1991, a component of the pedicellarial toxin of the urchin, *Toxopneustes pileolus* was isolated and purified by Nakagawa *et al.* (1991). Because of its ability to contract isolated pig tracheal smooth muscle, it has been called contractin A. Its molecular weight is 17,700 with a total residue of 138 amino acids; the N-terminal amino acid is serine. The authors suggested that the isolated smooth muscle may be affected through activation of phospholipase c. Takei *et al.* (1991) isolated a biotoxic substance, PII, from crude extracts of the pedicellariae of *T. pileolus* by fractionation on Sephadex G-200. They obtained three protein fractions and demonstrated that fraction PII caused histamine release from rat peritoneal mast cells.

3. *Holothuroidea*

According to Halstead (1965-70) the toxicity of sea cucumbers has been known since the days of Aristotle. Cooper (1880) reported that some sea cucumbers can discharge tenacious filaments from their visceral cavity and that these may produce painful wounds in humans. Saville-Kent (1893) observed that eating the sea cucumber *Stichopus variegatus* can cause death, although the toxicity of holothurians was questioned in the earlier literature (Cleland, 1913; Clark 1921). It remained to the observations of Yamanouchi (1929) to consider seriously the toxicity of these animals, although they appear to have been avoided as food for centuries. He observed that when fishes were placed in an aquarium to which aqueous extracts of *Holothuria vagabunda* were added the fish soon died. Subsequently, he obtained a toxic crystalline product, termed holothurin, and found it present in 24 of the 27 species of sea cucumbers examined (Yamanouchi, 1942, 1955).

Nigrelli (1952) observed that the toxic substance (holothurin) extracted from the Bahamian sea cucumber *Actinopyga agassizi* was composed of 60 percent glycosides and pigments, 30 percent salts, polypeptides and free amino acids, 5–10 percent insoluble protein and 1 percent cholesterol. The cholesterol-precipitated fraction, known as holothurin A, represented 60 percent of the crude holothurin and was given the empirical formula $C_{50-52}H_{81-85}O_{25-26}SNa$ and I.R. absorption at 1748 and 1629 cm^{-1}. Holothurin appeared to consist of at least four steroid aglycones bound individually to four molecules of monosaccharides; it showed no absorption in the ultraviolet region. It was probably a mixture of aglycone of approximately 26 to 28 carbon and four to five oxygen atoms, one molecule each of four different sugars and one molecule of sulfuric acid as a sodium salt. A provisional structure was published by these workers but was altered by Friess and Durant (1965) (Fig. 40). It resembled digitonin and other saponins in both its chemical and biological activities. The toxin had a deleterious effect on some sharks and was suggested as a shark repellent.

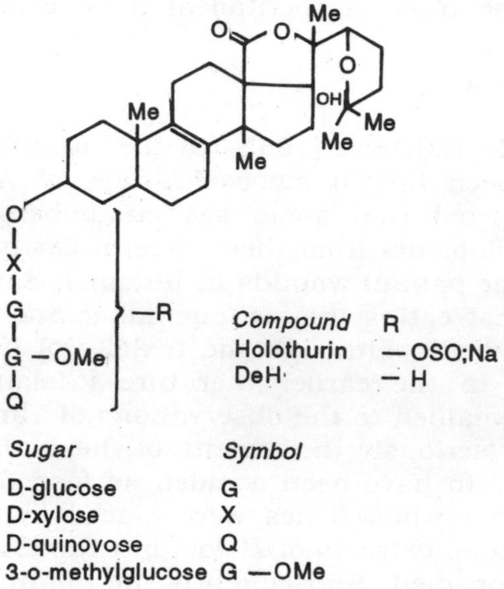

Figure 40. Provisional structure for holothurin A (Friess *et al.*, 1967).

Holothurin, 10 ppm, was found to be lethal to *Hydra*, the mollusc *Planorbis* and the annelid *Tubifex tubifex*. It had slightly greater hemolytic action than saponin, and stimulated hemopoieses in the bone marrow of winterized frogs (Nigrelli and Jakowska, 1960). Administration of C^{14}-labeled mevalonic acid into *S. badionotus* was incorporated into holothurin, which might indicate that the animal synthesizes holothurin *in vivo* (Murthy and Der Marderosian, 1973). In the mammalian phrenic nerve-diaphragm preparation, holothurin A produced a contracture in the muscle, followed by some relaxation, and a gradual decrease in the recorded amplitude of both the directly- and indirectly-elicited contractions, the latter decreasing at a slightly greater rate than the former. The intravenous LD_{50} in mice was approximately 9 mg/kg body weight (Friess *et al.*, 1960). In experiments in frogs, Thron and his colleagues (1963) demonstrated that holothurin

A produced an irreversible block and destruction of excitability on the single node of Ranvier in the sciatic nerve. The toxin did not produce any observable damage to the axonal walls or sheath. The poison did not exert a blocking action on the *in vitro* AChE-ACh system.

The major aglycone from the holothurin A of *Holothuria vagabunda*, named holothurigenin, was isolated by Matsumo and Yamanouchi (1961). Holothurigenin, $C_{30}H_{11}O_5$, had a melting point of 301°C, contained three hydroxyls, a five-membered lactone, and a heteroannular diene. On hydrolysis, Chanley *et al.* (1966) isolated two genins by repeated alumina column chromatography of a mixture of aglycones derived from the holothurin A of *Actinopyga agassizi*. The two genins accounted for 30 percent of the total aglycone and were identified by spectral and chemical techniques as 22.25-oxidoholothurinogenin and 17-desoxy-22.25-oxidoholothurinogen.

Yasumoto *et al.* (1967) isolated two saponins from the body walls of *Holothuria vagabunda* and *H. lubuca* by column chromatography on silic acid. The major saponin was obtained as needles, m.p. 225–226°C (dec.) and had an empirical formula of $C_{55}H_{99}O_{29}SNa$. Its sugar composition and other properties were identical with those of holothurin A isolated by Chanley *et al.* (1966). The other saponin, named holothurin B, was isolated as needles, m.p. 213–216°C (dec.). It had an empirical formula of $C_{45}H_{75}O_{20}SNa$ and I.R. bands at 1745 and 1640 cm^{-1}. On acid hydrolysis it was found to be a mixture of aglycones, sulphuric acid, and one mole each of D-quinovose and D-xylose. Matsuno and Iba (1966) also isolated holothurin B from *H. vagabunda*. Table 17 shows the chemical properties of holothurins A and B as published by Hashimoto (1979). Shimada (1969) isolated a toxin from the body wall of *Stichopus japonicus* which closely resembled the holothurins. He called the toxin holotoxin. It is a steroid glycoside which differs from holothurin in that it is not a sulfate and has unique antifungal properties. It is similar to holothurin in its infrared spectrum, showing bands at 1745 and 1640 cm^{-1}, indicative of a five-membered ring lactone and one double bond, respectively.

Habermehl and Volkwein (1968, 1970, 1971) investigated the toxins from the Cuvierian organ and body wall of a number of species and summarized the findings in Table 18.

Table 17. Properties of holothurins A and B.

Property	Holothurin A	Holothurin B
Melting point	224–226°C (dec.)	213–236°C (dec.)
Specific rotation $(\alpha)_D^{20}$	–14.4°	–7.3°
Hemolytic index	187 000	125 000
Empirical formula	$C_{55}H_{99}O_{29}SNa$	$C_{45}H_{75}O_{20}SNa$
U.V. absorption	None	None
I.R. absorption	1745, 1640 cm^{-1}	1745, 1640 cm^{-1}
Sulphate group	One mole	One mole
Sugar	3-O-Methyl-D-glucose, D-Quinovose, D-Xylose D-Glucose (1:1:1:1)	D-Quinovose, D-Xylose (1:1)
Ability to form adduct with cholesterol	+	
Aglycone	Mixture	Genin-I, II, III

From Hashimoto, 1979.

According to the authors, these compounds are glycosides of tetracyclic triterpenes which are derivatives of lanosterol and were the first glycoside triterpenes derived from animals. In their 1971 paper, Habermehl and Volkwein proposed a new nomenclature for the toxins in an attempt to bring order to the classification difficulties presented by gimmick or trivial names.

Kitagawa *et al.* (1974, 1976), working with Shimda's holotoxin, found it was composed of three saponins, holotoxin A, B and C. All three substances on Hydrolysis yielded two genins, D-xylose, D-quinovose, 3-O-methyl-D-glucose and D-glucose. Their empirical formulas were $C_{57}H_{94}O_{27}H_2O$ (holotoxin A) and $C_{65}H_{104}O_{33}H_2O$ (holotoxin B). Other substances isolated from the holothurins include stichoposides (Anisimov *et al.*, 1973), cucumarioside C, $C_{30}H_{44}O_5$ (Elyakov and Peretolchin, 1970), and thelothurins A and B (Kelecom *et al.*, 1976). Anisimov *et al.* (1974) demonstrated that cucumarioside C, extracted from *Cucumaria fraudatrix*, produced changes in sea urchin embryo development leading to alterations in biopolymer syntheses. It differed from the stichoposides in that it affected sea urchin eggs at various stages of development, whereas stichoposides uniformly arrested egg division.

Holothurin has been shown to have hemolytic and cytolytic properties, among others. It is considered to be one

Table 18. Occurrence of toxins in Holothurioideae spp.

Species	Substance No.										
	1	2	3	4	5	6	7	8	9	10–17	18/19
Actinopyga agassizi	0	0	0	0	—	—	—	—	—	0	—
Halodeima grisea	0	—	+	—	—	—	—	—	—	—	—
Holothuria tubulosa	+	+	—	—	—	—	—	—	—	—	—
Holothuria forskåli	0 / +	0 / +	—	—	—	—	—	0	—	—	—
Holothuria poli	+	+	+	—	+	—	—	+	+	—	—
Bohadschia koellikeri	—	—	—	—	—	+	+	+	+	—	—
Stichopus japonicus	—	—	—	—	—	—	—	—	—	—	+

+, isolated from body wall; 0, isolated from Cuvierian organs.
From Habermehl and Volkwein, 1971. References in original table.

of the most potent saponin hemolysins known. In some concentrations it is lethal to animals and plants, inhibits the growth of certain protozoa, modifies the normal development of sea urchin eggs, retards pupation in the fruit fly, and inhibits regeneration processes in planariae. Some of its properties are shown in Table 19.

The effects of various preparations of holothurin on the peripheral nervous system were the object of a number of studies by Friess *et al.* (1959, 1960) and Thron *et al.* (1963). These workers demonstrated that holothurin in concentrations of 9.8×10^{-3} M caused a decrease in the height of the propagated nerve potential without reducing the conduction velocity in the frog desheathed sciatic nerve. This change was concentration-dependent and independent of pH, at least between pH 7.6 and 8.1 at 1.95×10^{-3} M. It was irreversible. A similar change was produced in the single fiber-single node of Ranvier preparation. In concentrations of $2.5 \times 10^{-5} - 1.0 \times 10^{-3}$ M, the toxin produced a diminution of the action current with a concomitant rise in the stimulation threshold. This change was also irreversible and was unaffected by physostigmine in concentrations 100 or 400 times greater than the holothurin concentrations.

Table 19. Effects of holothurin on animals and plants

Organism	Holothurin (ppm)	Effects
Protozoa		
Euglena gracilis	66	Growth inhibition
Ochiomonas malhamensis	22	Growth inhibition
Tetrahymena pyriformis	22	Growth inhibition
Amoeba proteus	100	Lethal
Paramecium caudatum	10	Lethal
Coelenterata		
Hydra (brown)	10	Lethal
Platyhelminthes		
Dugesia tigrina	0.1–100	Inhibition of regenerative processes; lethal
Nemathelminthes		
Nematode (Free living)	100	lethal
Mollusca		
Planorbis sp.	1–10	Lethal
Annelida		
Tubifex tubifex	10	Lethal
Arthropoda		
Insect, *Drosophila melanogaster*	1000	Retardation of pupation
Echinodermata		
Arbacia punctulata, eggs	0.001–1000	Developmental changes
Chordata		
Osteichthyes, *cyprinodon baconi*	1–100	Lethal
Carapus bermudensis	< 1–1	Lethal
Amphibia, *Rana pipiens*	10 (mg)	Lethal (mg/animal); hemolysis and increased hemopoiesis
Mammalia, mouse	9 (mg)	LD_{50} (IV, mg/kg)
	10 (mg)	MLD (IP, mg/kg)
Plants		
Water cress, root hairs	1000	suppression of development
Onion, root tips	1000	Necrosis; lethal

From Alender and Russell, 1966. References in original table.

On histological examination of the single fiber-single node preparation following incubation with 8.7×10^{-6} M holothurin, no change was observed at the lowest concentration that produced the decrease in action current. Thron *et al.* (1964), however, found that in approximately 80 percent of the preparations studied the loss of nodal excitation caused by this same concentration was accompanied by loss in basophilic, macromolecular material from the axoplasm in and near the node of Ranvier.

In the rat phrenic nerve-diaphragm preparation, Fries *et al.* (1959, 1960) and Thron *et al.* (1964) found that holothurin A, at a concentration of 1×10^{-5} M and above, produced an irreversible depression in the height of both the directly and indirectly elicited contractions, the latter decreasing at a slightly greater rate than the former. These concentrations also produced an acute contracture of short duration of the muscle which was independent of electrical stimulation. A contracture was also observed following topical application of holothurin A on frog muscle (Thron *et al.*, 1963). If the nerve-muscle preparation was pretreated with physostigmine at concentrations of the 10^{-10}–10^{-9} M range, and then challenged with holothurin A, much of the toxin's ability to block conduction at the junction was thwarted. This "protective" effect of physostigmine was lost when concentrations were increased to 7×10^{-9} M and higher.

Friess *et al.* (1967) showed that the crystalline holothurin A produced a potency and degree of irreversibility of the agonistic actions in the mammalian phrenic nerve-diaphragm preparation that were dependent on the presence of the charged sulfate residue of the toxin. The blocking actions of a selective desulfated derivative were, for the most part, reversible on washing, in contrast to the irreversible changes produced by the parent toxin. Lethality determinations in mice indicated that the signs preceding death were the same for the two substances, but that the parent toxin provoked these signs at a very much accelerated rate. Subsequently, Friess *et al.* (1970) studied the action of holothurin A elaborated by *Actinopyga agassizi* and its neutral desulfated derivative on the cat superior cervical ganglia. Both saponins produced irreversible inactivation of the ganglion; the holothurin A was the more active. Using the rat phrenic nerve-diaphragm preparation, Friess *et al.* (1968) compared holothurin A from the sea cucumber *A. agassizi*, holothurin B from *H. vagabunda* and *H. lubrica*, and asterosaponins A and B from the sea star *Asterias amurensis*. Table 20 shows the structural properties of these saponins.

Summarizing the various studies of Friess *et al.*, it can be noted that the most obvious functional similarity from data with the cervical ganglia of the cat, the peripheral neuromuscular junction of the rat and the medullated nerve nodes of the frog was the possession of cholinergic subsystems at some anatomical level, chiefly within the excitable membranes of conducting and junctional structures. These workers felt a common target for holothurin A action was "the cholinergic receptor population triggered by acetylcholine ion (ACh⁺) in the associated hydrolase enzyme AChE, or in the enzyme choline acetylase responsible for resynthesis of ACh⁺."

DeGroof and Narahashi (1974) studied the action of holothurin A on the electrical properties of the squid axon. They found that external application of the toxin to the axon produced an irreversible depolarization of the membrane to nearly zero potential, while internal perfusion caused a biphasic depolarization of the membrane. The time course of the depolarization was much shorter with the internal application than with the external application. They postulated that their results demonstrated an increase in resting sodium permeability as one of the mechanisms underlying the depolarization by holothurin A.

Jayasree *et al.* (1989) found that the saponin levels in the Cuvierian glands of the Indian holothurian, *Holothuria leucospilota* were high throughout the year, while this was not true of the levels in the body wall and viscera. Temperature also appeared to influence the amount of saponin concentrated in the Cuvierian glands but not in the body wall. When studied on shrimp, it was found that lactate dehydrogenase was increased, possibly indicating diminished cellular oxidation; glutamate dehydrogenase was increased, possibly indicating the role of glutamic acid in amino acid metabolism; succinate dehydrogenase and maltate dehydrogenase were inhibited, possibly indicating impairment of respiratory metabolism and cellular oxidative process; alanine and alkaline amino transferase activity were increased, possibly due to increased protein catabolism; and acid phosphatase activity and decreased alkaline phosphatase were found, thought to be due to necrotic changes in cells and transphosphorylation, respectively (Ashok Kumar, *et al.* 1991). Rat brain Na⁺–K⁺ATPase is inhibited by the triterpene glycosides of holothurians (Gorshkova *et al.*, 1989).

Table 20. Structural properties of purified Echinoderm saponins.

Saponin	Animal source	Presence of -OSO$_3$-Na$^+$ group	Sugars per molecule	Empirical formula	Molecular weight (average)
Holothurin A	*Actinopyga agassizi* Selenka	Yes	1 D-Xylose 1 D-Glucose 3-O-Methyl-D-glucose	$C_{50-52}H_{81-85}O_{25-26}SNa$	1159
	H. vagabunda	Yes		$C_{55}H_{99}O_{29}SNa$	1279
	H. lubrica	Yes	1 D-Quinovose		
Holothurin B	*H. vagabunda*	Yes	1 D-Xylose 1 D-Quinovose	$C_{45}H_{75}O_{20}SNa$	991
	H. lubrica	Yes			
Asterosaponin A	*Asterias amurensis* Lütken	Yes	2 D-Quinovose 2 D-Fucose	$C_{50-52}H_{88-90}O_{25-26}SNa$	1141
Asterosaponin B	*Asterias amurensis* Lütken	Yes	2 D-Quinovose 1 D-Fucose 1 D-Xylose 1-D-Galactose	$C_{56}H_{101}O_{32}SNa$	1341

From Friess *et al.* 1967. References in original table.

The triterpene glycosides of the *Holothuria* are membrane-active substances. They appear to alter the membrane permeability for K+ ions as well the u.v.-absorbing properties of erythrocytes, eucaryotic cells and liposomes formed from phospholipids and cholesterol (Dettbarn, 1965; Anisimov *et al.*, 1978; Kuznetsova *et al.*, 1983). This effect is thought to be related to the formation of glycosidesterol complexes (Nigrelli and Jakowska, 1960).

C. Clinical Problem

The late Dr. Carl Hubbs of the Scripps Institute of Oceanography called one of us (FER) in consultation following an accident in 1965 involving a student working with *Acanthaster planci*, who had inadvertently slipped and fallen, landing forcibly with his left hand impaled upon the sea star. Twenty minutes after the injury the patient had intense pain over the palm of the left hand, "shooting pains" up the volar aspect of the forearm, weakness, nausea, vertigo and tingling in the fingertips. There were at least ten puncture wounds over the hand, some of them bleeding freely. It was not certain if some of the spines had broken off in the wound. I suggested that the patient put his hand in cold vinegar/water and be admitted to the hospital emergency room. On arrival there 15 minutes later, the pain was less intense and the nausea had somewhat subsided. Unfortunately, the patient was given 100 mg of meperidine hydrochloride and 5 minutes later he was vomiting. (The vomiting was probably due to the medication, and most of the other symptoms to hyperventilation.) The patient's hand was placed in cold vinegar/water and occasional aluminum acetate soaks over the next two days and all symptoms and signs, including the mild edema, slowly resolved. Several broken spines were removed from the puncture wounds. Four days following the accident the patient complained of burning and itching over the left palm. Examination revealed a scaly, erythematous dermatitis. Topical corticosteroids were used but two days later the patient had to be placed on systemic corticosteroid therapy for the allergic dermatitis, which cleared in six days.

A second episode involving *A. planci* was related to FER by W.L. Orris in 1974. The patient had immediate, severe, burning pain and localized edema. These responded to aluminum acetate soaks and corticosteroids. According to Endean (1964), the puncture wounds produced by *Asthenosoma periculosum* give rise to immediate and sometimes acute pain but few other symptoms or signs. The discharge from *Marthasterias glacialis* is said to cause edema of the lips (Giunio, 1948). Halstead (1980) notes that contact with venomous sea stars causes "a painful wound accompanied by redness, swelling, numbness, and possible paralysis." Allergic dermatitis can occur following extensive contact with sea stars.

Although few people appear to enjoy eating sea stars, it is known that cats fed on *Solaster papposus* subsequently succumb (Parker, 1881), and that marine animals exposed to the release of the saponin or "saponin-like toxin" in the water will react in avoidance or, as in the case of a confined area, will die.

Stings by pedicellariae of certain sea urchins are well-documented (Russell, 1965a; Halstead, 1965–70; Cleland and Southcott, 1965; Sutherland, 1983; Hashimoto, 1979; Russell, 1984; Halstead, 1988). Fujiwara (1935) experienced severe pain, syncope, respiratory distress, partial paralysis of the lips, tongue and eyelids, and weakness of the muscles of phonation and extremities following stinging by seven or eight pedicellariae from *Toxopneustes pileolus*. Mortensen (1928–51) experienced severe pain of several hours duration at the site of a stinging by *Tripneustes gratilla*. The sting of a single globiferous pedicellaria from *T. gratilla* was found to be equal in pain severity to that experienced following a bee sting. Swelling appeared around the puncture wounds within minutes of the stinging, and a red wheal of a centimeter in diameter soon developed. Subsequent stingings during the following two-year period resulted in a more severe reaction. In one instance, the wheal was 12 cm in diameter and persisted for eight hours. In none of these cases were there any systemic manifestations (Alender and Russell, 1966; Russell, 1971b). It might be concluded that pedicellarial stings give rise to immediate pain, localized swelling and redness, and an aching sensation in the involved part. Other findings might include those described by Fujiwara (1935) and reviewed by Halstead (1965–70).

As previously noted, the secondary spines of *Echinothrix calamaris, E. diadema, Asthenosoma varium, A. thetidis* and the primary oral spines of *Phormosoma bursarium* are said to have a venom gland and are capable of envenomation (Alender and Russell, 1966). However, case reports on verified stingings are almost nonexistent, and in the several known to the authors, it is not possible to decide whether the pain, "dizziness" and minimal localized swelling were due to a venom or the effects of a simple puncture wound and manifestations complicated by hyperventilation.

The primary spines of almost 50 species of sea urchins have been implicated in traumatic injuries to man. Urchins of the family Diadematidae are particularly troublesome because of their spine length and fragility. When these break off in a puncture wound they can be difficult to find and remove. One of us (FER) has attended injuries in which it was necessary to remove more than a dozen broken spine tips from flesh. With some species there is no giveaway dark color around the puncture wound, which makes finding the broken spines difficult. Although the fragments of some spines will dissolve in human tissues and cause no difficulties, others can give rise to granulomatous reactions, occasionally requiring surgical removal (Fig. 41). Still others may migrate through the foot or hand without causing complications. While working in our laboratory, Dr. C.B. Alender had the tip of a spine from *T. gratilla* pass through his left hand. It had dissolved slightly but was still recognizable. It required two years to migrate through his hand. Occasionally, spines will lodge against a nerve or bone and cause complications requiring surgery. Secondary infections from spine injuries are relatively rare.

Figure 42 is a xerogram on a patient, showing several spine tips in the tissues. The xerogram (mammogram) is the best method of locating such fragments. The spine pieces were removed surgically. In freeing any such sea urchin spines, great care must be taken not to fragment them during removal. It may be necessary to enlarge the removal incision more than would be necessary for a simple splinter, glass or other object, so that it can be more easily freed and worked out.

The ovaries of sea urchins have long been known to be toxic and sometimes lethal to humans (Loisel, 1903, 1904).

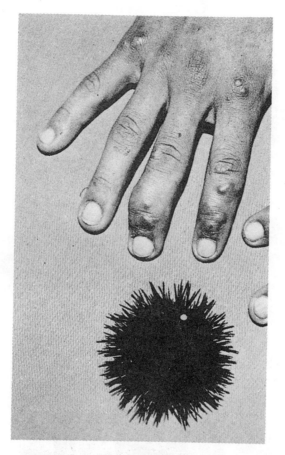

Figure 41. Sea urchin spines broken off in hand of diver with resulting granulomatous reaction (Courtesy: R.V. Southcott).

The gonads of *Paracentrotus lividus*, *Tripneustes ventricosus* and *Centrechinus antillarium* have been noted to be poisonous (Halstead, 1980). The eating of many species of sea urchin eggs is common in many parts of the Indian Ocean. The ingestion of certain sea cucumbers is also common in the South Pacific, Philippines, Japan, China and other areas of the Orient and Near East. The most frequently implicated species are *Holothuria atra*, *Holothuria axiologa*, *Stichopus variegatus* and *Thelenota ananas* (Yamanouchi, personal communication, 1958). In most cases, the clinical manifestations are of short duration and without serious sequelae. There may be some nausea, and epigastric pain, followed by vomiting, but these generally pass quickly. In the more

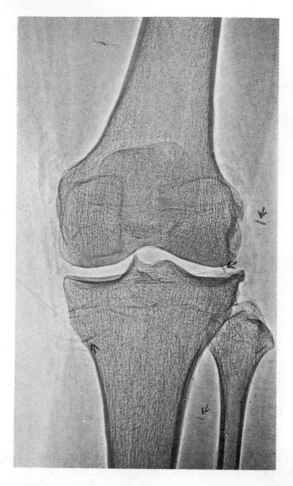

Figure 42. Xerogram of left knee, showing broken *Diadema* sp. spines embedded in tissues of left knee of diver. Demonstration of the spines allows for easy removal.

serious cases, severe dehydration may occur. Treatment is symptomatic, with early gastric lavage, plenty of fluids, mild sedation, and respiratory assistance, if necessary.

Handling some holothurian species may cause pruritus, redness of the hands and some burning pain. In a patient seen by one of us (FER) on Okinawa, a severe *venenata* dermatitis developed in a diver who had been collecting *Holothuria* sp. for many years. He was advised to cease such collecting and was placed on a 2 percent corticosteroid cream and parenteral corticosteroid. The medications were continued for several weeks, along with twice daily Burows soaks. The lesions slowly healed without complications. Unfortunately, several weeks after the dermatitis had cleared, the patient renewed his collecting and again experienced mild pruritus with acute swelling of hands, along with some weakness and dyspnea. When seen by Dr. S. Sato on Okinawa, he was again treated with a parenteral corticosteroid and strongly advised to cease collecting sea cucumbers. In our last communication with Dr. Sato, we were informed that the patient had given up collecting sea cucumbers and had now turned to collecting sea snakes, out of the frying pan into the fire. Acute conjunctivitis has been reported in divers collecting holothurians in waters polluted with the tissue discharge of the Cuvierian organs (Russell, 1965a).

MOLLUSCA

A. Introduction

The phylum Mollusca includes slugs, snails, whelks, clams, mussels, oysters, scallops, octopods, squid and nautili. They are distinguished by their bilateral symmetry, unsegmentation and a mantle which often secretes a calcareous shell. They have a ventral muscular foot that in the different classes is used for locomotion, digging or seizing prey. Their coelom is reduced, they have an open circulatory system and radula or tongue-like organ, absent in the bivalves. Some species have jaws; respiration is usually by means of gills. There are approximately 80,000 species of which about 100 have been implicated in poisonings to man, or are known to be poisonous or venomous under certain circumstances. The majority of toxic species are found in three of the five classes: Gastropoda, Cephalopoda and Pelecypoda.

The Gastropoda include the dangerous cones of the genus *Conus*, of which there are approximately 400 species. Almost all these are confined to tropical or subtropical seas and oceans. They are usually found in shallow waters along reefs, although some of the more dangerous species are found on sandy bottoms. They range in length up to approximately 25 cm. Other toxic gastropods are found in species of *Aplysia, Creis, Haliotis, Livona, Murex, Thais* and *Neptunea*. Halstead (1980) estimates that there are 33,000 living species of gastropods. In the cephalopods, some cuttlefishes, squids, nautili and octopods are known to be venomous, and several poisonous squids are known. Among the poisonous pelecypods: the bivalves, clams, mussels and oysters, are the Japanese callista, *Callista brevisiphonata*; the giant clam, *Tridacna maxima*, common to some parts of the seas of the South Pacific; the oyster, *Crassostrea gigas*, and the shellfishes, *Dosinia japonica* and *Tupes semidecussatu*. Of course, PSP can be found in various members throughout the Mollusca.

B. Venom Apparatus

The poisonousness of the clams, mussels and certain other Mollusca can be attributed to PSP or related protistan poisons. This subject is reviewed in the chapter on Protista. In *Conus*, the venom apparatus is said to be homologous with the unpaired gland of Leiblein found in certain of the higher gastropods (Fig. 43). It serves as an offensive weapon for food acquisition and, to a lesser extent, as a defensive weapon against predators. It consists of muscular bulb, a long coiled venom duct, the radula, the radula sheath and the radular teeth. The muscular pharynx and extensible proboscis are considered to be accessory organs.

The venom apparatus in *Conus* lies on the dorsal side of the animal in a cavity posterior to the rostrum. The venom is thought to be secreted in the venom duct and forced under pressure exerted by the duct into the radula and then into the lumen of the radular teeth. According to Kohn *et al.* (1960), the length of the venom duct may be 15 times greater than the straight line distance, and ducts may be 5–10 cm in length, depending on the species. Although the venom bulb has sometimes been implicated as

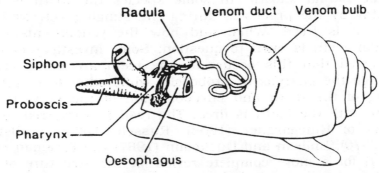

Figure 43. Schematic representation of relationship of venom apparatus in *Conus* (Russell, 1984).

the source of venom production, the works of Hermitte (1946), Hinegardner (1957) and Endean and Duchemin (1967) certainly indicated that the venom is produced along the length of the duct, with the posterior duct area contributing the most toxic portion. Venom may be found in the lumen of the bulb (Kohn *et al.*, 1960) but Endean and Duchemin (1967) suggest that this might be reflux from the posterior region of the venom duct and that the "sac-like cytoplasmic outgrowths containing spherules occasionally observed in the bulb epithelium are probably ruptured or separated from the epithelium when the musculature of the bulb contracts." The thick musculature of the venom bulb, and its ability to forcibly squeeze the bulb, probably account for the movement of venom along the duct and into the pharynx.

The radula is a Y-shaped organ lying anterior to the esophagus-stomach and opening into the pharynx just anterior to the entrance of the venom duct. It produces the radular teeth. The organ is divided into three sections. The largest section, which overlies the esophagus-stomach, may contain as many as 30 teeth in various states of development. The short arm of the gland attaches to the pharynx and contains most of the mature teeth. The third section is known as the ligament sac.

The radular teeth are passed from the radula into the pharynx, then into the proboscis. There they are thrust by the proboscis into the prey during the stinging act (Fig. 44). The radular teeth are needle-like, 1 to 10 mm in length and almost transparent. They vary in size and shape depending on the species. A ligament is attached to the base of each tooth and serves as a means of fixation while the tooth is in the radula sheath. In some species the tooth is held forcibly by the proboscis during the stinging act, while in other it is freed. When and how the venom enters the radular teeth is open to question. Some investigators have suggested that this occurs as the teeth are being moved through the pharynx or proboscis, while other feel that the filling of the tooth and envenomation of the prey do not occur until the tooth is fired. The reader is referred to the works of Hinegardner (1958), Russell (1965a), Halstead (1965–70), Endean and Duchemin (1967),and Freeman *et al.* (1974) for a more complete review of the structure of the venom apparatus of *Conus*. The venom apparatus of the

Figure 44. Cone firing radular tooth from proboscis into a sea snail.

Indian cone, *Conus amadis*, has been described by Kasinathan *et al.* (1989).

In the octopus, the venom apparatus is an integral part of the animal's digestive system. The secretions serve in prey capture and digestive function, in some ways akin to the venoms of snakes. The apparatus consists of paired posterior salivary glands, two short (salivary) ducts which join them with the common salivary duct, paired anterior salivary glands and their ducts, the buccal mass and the mandibles, or beak (Fig. 45). The two paired salivary glands may differ in their size, structure and function. The common salivary duct opens into the sub-radular organ anterior to the tongue. The paired ducts from the anterior salivary glands open into the posterior pharynx. The buccal mass, or pharynx, is a muscular complex with two powerful horny jaws like an inverted parrot's beak.

As long ago as 1909, Isgrove found venom glands of *Eledone* contained glandular secretory tubules embedded in

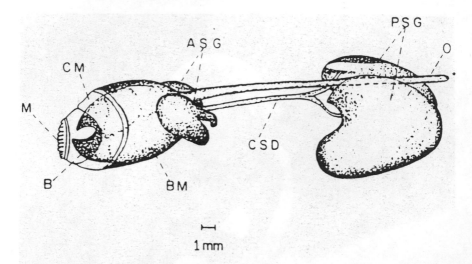

Figure 45. Diagrammatic sketch of the venom apparatus of an octopus. ASG, anterior salivary glands; B, beak; BM, buccal mass; CM, circular muscle; CSD, common salivary duct; O, oesophagus; M, mouth; PSG, posterior salivary glands (Russell, 1984).

a stroma of connective tissue. The secretory cells were columnar with a basally oriented nucleus. The secretion was formed in globules in the anterior part of these cells and passed into the lumen of the tubules. He described the secretion as a kind of mucus, containing "no ferment whatever". F.R Modglin (see Halstead, 1988) examined the venom glands of *Octopus rubescens* and found that in the anterior salivary gland the glandular elements were lined by a simple columnar epithelium containing either peripherally or basally placed dark granules in the cytoplasm and having vesicular nuclei. He was unable to find duct-like structures. In some areas the lumina were dilated and contained a web-like acellular coagulum. The posterior glands consisted of fairly close-packed acini of glandular tissue that were mucoid in nature, having large glandular cells with clear swollen cytoplasm and an abundance of secretory granules. Modglin found numerous duct-like structures lined by tall, simple, columnar epithelium. These ducts were surrounded by a hyaline band similar to a basement membrane. The varying shapes of the acini and tubules suggested a convoluted arrangement and he concluded that the salivary glands resembled the mucous glands of vertebrates and have a merocrine type of secretion.

Gennaro *et al.* (1965) studied the salivary glands of *O. vulgaris* and found that the posterior glands consisted of branched, convoluted tubules contained within a mesenchymatous mass. Acinar spaces were lined with secretory cells varying from cuboidal to columnar and containing secretory droplets. Within the acini were masses of pale protein material. Matus (1971), examining the posterior salivary gland of *Eledone cirrosa* and *O. vulgaris*, described two distinct types of epithelia lining the tubules. One type, which lined the wide lumen tubules, had a secretory function. The other did not. Gibbs and Greenaway (1978) studied the posterior salivary glands of *Hapalochlaena maculosa* and *O. australis*, the latter being a non-venomous species, and found that in *H. maculosa* the gland was composed of compound tubuloacinar exocrine structures and possessed mucous secretory cells. In *O. australis*, the gland was a mixture of tubular and tubuloacinar elements. They suggested that the transport of the saliva in *H. maculosa* is accomplished by muscular contraction of the gland by ciliary action and that the saliva of *O. australis* had a purely digestive function. They also suggested that there are three types of octopods, and that they represent successive evolutionary states, as far as the development of the posterior salivary glands are concerned. The types would range from those where saliva has a purely digestive function to those which develop highly specialized toxins.

C.   Chemistry and Pharmacology

The earlier studies on the venom of *Conus* indicated that the venom was white, gray, yellow or black, depending on the species, viscous, and contained soluble proteins, peptides, low molecular weight compounds and insoluble granules. The granules were rich in Cys-containing proteins and have been suggested as the processed precursors of the soluble peptides. The venom pH ranged from 7.4–8.3. Early studies indicated that the active principle was non-dialyzable, and its toxicity was reduced by heating or incubation with trypsin. Extracts of the venom ducts yielded homarine, gamma butyrobetaine, N-methyl-pyridinium, several amines (possible indole amines), 5-hydroxtryptamine, lipoproteins and carbohydrates. The lethal fraction was thought to be a protein or bound to a protein (Kohn *et al.*, 1960; Whysner and Saunders, 1966). Kohn also suggested that the various species of Conidae could be divided into those that were vermivorous, molluscivorous or piscivorous and suggested that in mammals the primary pharmacological deficit might be in neuromuscular transmission. The Endean group, using the venom from *C. geographus* (Whyte and Endean, 1962), *C. magus* (Endean and Izatt, 1965), and *C. striatus* (Endean *et al.*, 1967) were not able to demonstrate a specific neuromuscular change and attributed the effect to action directly on the muscle.

Endean and Rudkin (1963–1965) and Endean *et al.* (1967) demonstrated that of 37 species studied, the effects of the venoms were indeed directly related to the prey hunted. They concluded that only the piscivorous Conidae were capable of serious injuries to man. Subsequently, the Endean group found that the venoms of *Conus magus* and *C. striatus* contained conspicuous ellipsoidal granules possessing a peripheral film of polysaccharide, a sheath in which protein and lipid could be identified, and a core containing three indolyl derivatives. The average size of these granules appeared to increase as they moved from the anterior to the posterior end of the venom duct. In venom taken from the posterior end of the duct, the granules had a mean long axis of 9 µm and a mean short axis of 4 µm.

In 1974, Freeman *et al.* found that the amount of venom in the ducts of *C. striatus* was sufficient to immobilize the small fish on which it preys but not sufficient to cause serious injury to a human. They found a toxin having a molecular weight of over 10,000 with a low lethal index as compared to other *Conus* species. The toxin caused ataxia, depressed respiration leading to apnea and cardiac arrest in mammals, precipitated a block in the compound action potential of the isolated toad sciatic nerve, blocked both the directly and indirectly elicited contractions of a mammalian nerve-muscle preparation, and markedly depressed the amplitude of intracellular recorded action potential in the rat diaphragm.

Various early authors reported that the active fraction of *Conus* was a protein with a molecular weight of about 10,000, or a low molecular weight substance. Proteases were found in the anterior part of the venom duct of a number of species but not in the posterior part (Marsh,1971). Table 21 shows some of the venomous species of *Conus* and the kinds of animals in which they appear to exert their deleterious effects.

During the past decade numerous studies have been carried out on the toxic fraction of *Conus* venom, first labeled conotoxin. Its principal role is to quickly immobilize the prey so that it can be engulfed and digested. It is now known that conotoxin is a group of small polypeptide toxins (Table 22). The most extensively studied are α-conotoxin, μ-conotoxin, ω-conotoxin, etc. Some of these have been produced synthetically, particularly for ion-channel studies. Different, but not always accepted abbreviations, have been given to the various species toxins: C_f = *C. californicus*; EB = *C. eburneus*; F = *C. fulmen*; G = *C. geographus*; M = *C. magnus*; Mm = *C. marmoreus*; O = *C. obscurus*; Qc = *C. quercinus*; Sp = *C. spurius*; S = *C. striatus*; Ts = *C. tissulatus*; Tx = *C. textile*; T = *C. tulipa*.

Many workers have used Roman letters to indicate a particular sequence variant, that is, GIIIA, GIIIB, etc. Gray *et al.* (1988) have suggested that natural variants and synthetic analogues be named by IUPAC rules. For example, Pro[7]conotoxin GIIIA:D-tyr, conotoxin GI. For anyone other than a molecularpharmacobiochemicoconographer the system might be a bit laborious, but the reader is referred to the Gray, *et al.* (1988) paper for the need of a system, or systems, of nomenclature. With respect to the physiopharmacology, the problem of variables appears more clear. Those toxins seemingly associated with inhibiting the acetylcholine receptor at the postsynaptic terminus have been identified as α; those which block the muscle voltage sensitive sodium channel as μ; and those that selectively block Ca^{2+} channels in the terminals of perivascular nerves, reducing release but not affecting the contractible property of sympathetic transmitter as ω. Table 21 shows some representative conotoxins found in one cone shell, *C. geographus*.

α-conotoxin is the smallest of the paralytic peptides of *Conus* venom. It has 13–15 residues, is basic, and has two disulfide bonds. The toxin has been further fractionated into GI and GII (GIA). As previously noted, α-conotoxin blocks the nicotinic acetylcholine receptor in skeletal muscle

Table 21. Some venomous species of *Conus*.

Species	Toxic to Man	Mice	Paralysis in Fishes	Molluscs	Polychaetes
PISCIVOROUS					
C. catus Hwass	X	X	X	O	O
C. geographus L.	X	X	X	O	O
C. obscurus Sowerby	X				
C. magus L.		X	X	+	O
C. stercusmuscarius L		X	X	+	O
C. striatus L.	X	X	X	+	O
C. tulipa L.	X	X	X	O	O
MOLLUSCIVOROUS					
C. ammiralis L.		O	O	X	?
C. aulicus L.	X?	X?	O	X	O
C. episcopus Hwass		O	O	X	O
*C. marmoreus L.	X?	O	O	X	O
C. omaria Hwass	X?	O	O	X	O
C. textilis L.	X?	X	X	X	X
*C. tigrinus Sowerby		O	O	X	O
VERMIVOROUS					
C. arenatus Hwass		O	N	O	X
C. eburneus Bruguière		N	N+	X	X
C. emaciatus Reeve		O	O	O	X
C. flavidus Lamarck		O	O	O	X
C. figulinus L.		O	O	+	X
C. imperialis L.	X	O	O	O	X
C. leopardus (Röding)	X		O	O	X
C. lividus·Hwass	X	N	O	O	X
C. miles L.		O	O	O	X
C. millepunctatus Lamarck		O	O	O	X
C. omaria Hwass		O	O	O	X
C. planorbis Born	O	+	O	?	X
C. pulicarius Hwass	X	N	O	O	X
C. quercinus Solander	X	N	O	O	X
C. rattus Hwass		O	X	O	X
C. sponsalis Hwass	X	O	O	+	X
C. tessulatus Born		N	X	+	?
C. virgo L.		O	X	+	X

* Synonymous.
X Experimental or clinical evidence indicates toxicity.
+ May be lethal but does not produce paralysis.
? Possibly toxic, or evidence questioned or conflicting.
N Produces localized necrosis.
O Non-toxic, insofar as is known.
From Russell, 1965a.

Table 22. Representative conopeptides from *Conus geographus*.

Systematic name		Physiological target	Action	Bioassay activity			Abundance
				Fish ip	Mouse ip	Mouse ic	
α-Conotoxin	GI	Muscle; nAChR	Blocks	Lethal	Lethal	Lethal	Major
μ-Conotoxin	GIIIB	Muscle; v-a Na^+	Blocks	Lethal	Lethal	Lethal	Major
(Pro^7) μ-Conotoxin	GIIIB	Muscle; v-a Na^+	Blocks	Lethal	Lethal	Lethal	Minor
ω-Conotoxin	GVIA	Nerve; v-a Ca^{2+}	Blocks	Lethal	None	Shakes	Major
ω-Conotoxin	GVIIA	Nerve; v-a Ca^{2+}	Blocks	Lethal	None	Shakes	Moderate
lys-Conopressin	G	Vasopressin receptor	Activates	None	None	Scratches	Minor

Structures:

GI	Glu Cys Cys Asn Pro Ala Cys Gly Arg His Tyr Ser Cys*
GIIIB	Arg Asp Cys Cys Thr Hyp Hyp Arg Lys Cys Lys Asp Arg Arg Cys Lys Hyp Met Lys Cys Cys Ala*
(Pro^7) GIIIB	Arg Asp Cys Cys Thr Hyp Pro Arg Lys Cys Lys Asp Arg Arg Cys Lys Hyp Met Lys Cys Cys Ala*
GV	Gly Glu Gla Gla Leu Gln Gln Gla Asn Gln Gla Leu Ile Arg Gla Lys Ser Asn*
GVIA	Cys Lys Ser Hyp Gly Ser Ser Cys Ser Hyp Thr Ser Tyr Asn Cys Cys Arg Ser Cys Asn Hyp Try Thr Lys Arg Cys Tyr*
GVIIA	Cys Lys Ser Hyp Gly Thr Hyp Cys Ser Arg Gly Met Arg Asp Cys Cys Thr Ser Cys Leu Leu Tyr Ser Asn Lys Cys Arg Arg Tyr
Conopressin G	Cys Phe Ile Arg Asn Cys Pro Lys Gly*

Key to table and comments will be found in the original table (From Gray et al., 1988).

(Gray *et al.*, 1988). The μ-conotoxins are 22-peptide amides, strongly basic, and cross-linked by three disulfide bridges. The toxin blocks the response to the direct stimulation of vertebrate muscle.

The venom of *C. inscriptus* has a direct effect on smooth and cardiac muscle, is stable to 70°C but loses 50 percent of its activity at 50°C. It is stable at high and low pH (Nallathambi *et al.*, 1991). It appears that ω conotoxin inhibits contractions of stimulated noradrenergic nerves in mouse and rat anococcygens muscle, rabbit ear artery and the rat deferens. It also inhibits positive chronotropic responses to noradrenergic nerves. The inhibitory effect of conotoxin ω-on the Ca^{2+} influx required for excitation-secretion coupling in the autonomic nerve terminal is not absolute, and it is overcome by repeated stimulation or by raising the Ca^{2+} concentration (DeLuca *et al.*,1990)

The venom of *C. virgo* is a marked cardiac depressant not blocked by phentolamine but abolished in the presence of verapamil (Shanmuganandam *et al.*, 1991). The crude venom of *C. bayani* produces an anesthetic effect quicker than xylocaine and inhibits the acetylcholine-induced contraction in the frog rectus muscle (Kasinathan *et al.*,1991). Conotoxin GI, the synthetic peptide from *C. geographus*, is about 2.5 times more potent in its neuromuscular blocking activity than MI, the synthetic product of *C. magus*. The activity is reversed by neostigmine, indicating their specificity for nicotine receptors (Marshall and Harvey, 1990).

Extracts from the venom duct of the California cone, *C. californicus*, produce cytolytic activity towards sheep, goat, cow and chicken erythrocytes. The responsible cytolytic toxin(s) is ca. 19,000 daltons, has isoelectric values of 6.0, 6.7 and 7.3, and is antigenic (Kreger, 1989).

Rajendran and Kasinathan (1987) have studied crude venom extracts of *Conus amadis* from India on the fish *Mugil cephalus* and the crabs *Scylla serrata* and *Thalamitta cranota* and on the freshwater fish *Oreochromis mossambicus*. Kasinathan *et al.* (1989) demonstrated that a lethal dose of toxin of *C. amadis* produced an increase in the hematocrit and RBC count, and a decrease in serum protein, free amino acid and the blood glucose of an estuarine fish. Further studies showed that the decrease in protein was greater than that of lipid and carbohydrate in the tissues of gills, brain, liver and viscera. The brain was the most acutely affected. μ-conotoxin IIIa has been synthesized using solid-phase peptide synthesis employing 9-fluroenylmethoxy-carbonyl chemistry. The synthetic toxin has the same retention time on reverse-phase HPLC as the natural product, thus demonstrating its use for probing sodium channels (Becker *et al.*, 1989). The value of the conotoxins as probes for ion channel and receptors has been reviewed by Cruz (1989) and Hong and Chang (1989).

Some abalone are toxic to eat. Hashimoto (1979) notes, in particular, *Haliotis sieboldi*, *H. japonica* and *H. discus hannai*. The toxin is concentrated in the digestive gland or liver which can be distinguished by its blue-green pigment, evident during the spring of the year in Japan. It is thought that the pigment, pyropheophorbide *a*, originates from chlorophyll in the seaweed on which abalone feeds (Tsutsumi and Hashimoto, 1964).

Hashimoto *et al.* (1960) had observed that ingestion of the viscera of *Haliotis* caused dermatitis in cats and humans. On the basis of this observation they carried out experiments demonstrating the importance of photosensitization in the development of the dermatitis, suggested an assay method and concluded that the green pigment was a chlorophyll derivative. The photosensitization was found to be dependent upon the amount of abalone consumed, time after abalone ingestion, and light intensity. Interestingly, Hashimoto (1979) noted that the use of fluorescent pigment should be prohibited from foods, and that care should be taken to see that drugs are not transformed into fluorescent substances in the body.

The hypobranchial gland of *Murex* produces a secretion which at first is colorless or yellow but on exposure to sunlight becomes brilliant violet (Tyrian purple) and gives off a strong fetid odor. The gland also produces a toxic secretion. Dubois (1903) was perhaps the first to study this toxic secretion in the purple gland of *Murex brandaris*. Roaf and Nierenstein (1907) found that the toxin gave certain color reactions similar to those displayed by adrenalin, but Jullien (1940) demonstrated that the toxin was not adrenalin, but an ester of choline. Erspamer and Dordoni (1947) showed that the active substance was not acetylcholine. They name the compound murexine.

Subsequent studies indicated that two pharmacologically active substances were present. One of these was enteramine or 5-hydroxytryptamine (Erspamer and Asero, 1952) and the other was murexine. Elementary analysis of murexine by Erspamer and Benati (1953) gave the empirical formula $C_{11}H_{18}O_3N_3$. Further study showed that murexine had the structure of β-[imidazolyl-(4)]-acrylcholine. It was thus known as urocanylcholine:

$$
\begin{array}{c}
H \\
| \\
N\!-\!C\!-\!H \\
| \qquad\qquad\qquad\qquad OH \\
H\!-\!C \qquad\qquad\qquad\qquad | \\
\| \\
N\!-\!C\!-\!CH\!-\!CH\!-\!COOCH_3\!-\!CH_3\!-\!N\!-\!(CH_3)_3
\end{array}
$$

Blankenship *et al.* (1975) isolated murexine from the midgut of the sea hare *Aplysia californica*. Whittaker's group identified senecioylcholine in the hypobranchial gland of *Thais floridana* and acrylylcholine in *Buccinum undatum* (Whittaker and Wijesundera, 1952; Whittaker, 1960). The amount of these cholinesters in the hypobranchial gland was approximately 1–5 mg/g tissue. They are said to exhibit muscarine-like and nicotine-like activity and that they cause cardiovascular changes with hypotension, increased respiration and gastric motility, and some contraction of the frog rectus muscles and guinea pig ileum. The intravenous LD_{50} of murexine in mice was 8.1–8.7 mg/kg.

Roseghini isolated dihydromurexine from the hypobranchial gland of *Thais haemastoma*. It was found in high concentration, 15–17 mg/g tissue, in the yellowish median zone of the gland. Its action on the frog rectus muscle was ten times greater than murexine or acetylcholine. Hemingway (1978) demonstrated that the muricacean gastropod *Acanthina spirata* produced a paralytic substance with high acetylcholine content in both its hypobranchial and salivary/accessory/salivary gland complex. The toxin was thought to be a carboxylic ester of choline. The substances extracted from both glands were the same and were not identical to any of the cholinesters reported by Keyl *et al.* (1957).

Huang and Mir (1972) described a vasodilator and hypotensive agent in a salivary gland extract of the gastropod *T. haemastoma*. The extract produced behavioral changes in mice, followed by lethargy. When lethal doses were given, 43 mg fresh weight of gland/kg body weight, respiration first increased and then decreased and became shallow, and death ensued. The toxin produced bradycardia and a fall in blood pressure, which was partly blocked by atropine. In the isolated rabbit heart the extract produced a decrease in rate and contraction, with a fall in heart output. It also produced contraction of the guinea pig ileum and rabbit duodenum.

Bender *et al.* (1974) found that in addition to acetylcholine, other choline esters were present in the hypobranchial glands of *Acanthina spirata*, and that *Nucella (=Thais) emarginata* contained an N-methyl derivative of murexine, β-[4-(1 or 3-methyl)-imidazolyl] acrylcholine, which differed pharmacologically from murexine.

The salivary poison of the gastropod *Neptunea arthritica*, as well as certain other species sometimes eaten in the Orient, is tetramine ($C_4H_{12}N$). In 1960, it was suggested that histamine, choline and choline ester, also found in the salivary glands of this mollusc, act synergistically with the tetramine in producing the poisoning (Asano and Itoh, 1960) but more recent studies appear to indicate that there is no synergistic effects with the other main components, betaine and homarine (Anthoni *et al.*, 1989). Fänge (1960) noted that in *N. antigua* the tetramine was probably responsible for almost all of the biological activity of the salivary gland extract. He found that approximately 1 percent of the gland consisted of tetramine. Whittaker (1960) isolated an acetyl-choline-like substance from the gastropod *Buccinum undatum*. Infra-red comparison with a synthetic senecioyl-choline indicated that the active principle was an α,β-unsaturated derivative of isovalerylcholine. Tetramine has also been found in several other gastropods, one cephalopod and in the phyla Bryozoa and Cnidaria.

In the viscera of the ivory shell, *Babylonia japonica*, Hashimoto *et al.* (1967a) found a water soluble, slightly methanol and ethanol soluble, heat labile, dialyzable, ninhydrin positive, and Dragendorff and biuret negative toxin that had potent mydriatic activity. They used the mydriatic effect as an assay method in mice. Subsequently, a partially purified fraction, which was further purified on Sephadex G-25, was found to provoke mydriasis following subcutaneous injection into mice at the 0.02 mg/kg body weight level. The toxin was thought to be a polyfunctional, highly oxygenated complex bromo compound having a relatively low molecular weight (Shibota and Hashimoto, 1970).

In 1972, Kosuge *et al.* isolated a toxic factor in crystalline form from the mollusc *Babylonia japonica* and determined its molecular formula, $C_{25}H_{26}BrN_5O_{13} \cdot 7H_2O_{14}$, with a molecular weight of 810.53. Its structure was shown as:

The toxin was named *surugatoxin* (SGTX) after Suruga Bay, where the molluscs were taken. Another extract, *IS-toxin*, $(C_{22}H_{36}BrN_4O_{14})$ had previously been shown to produce blocking of the superior cervical ganglion in the cat (Hirayama *et al.*, 1970). Further studies of this showed that the toxin had a ganglion-blocking action, that it affected neither directly nor indirectly, elicited contractions in the mammalian phrenic nerve-diaphragm preparation, that it produced a prolonged hypotension in cats unaffected by atropine or spinal cord transection, and that the mode of anti-nicotinic action appeared different from that of hexamethonium and tetramethylammonium, and that it resembled that of mecamylamine.

As previously noted, the Pelecypoda: scallops, oysters and clams, are the principal transvectors of paralytic shellfish poisoning, discussed under Protista. The genera most often involved with PSP are *Mya*, *Mytilus*, *Modiolus*, *Protothaca*, *Spisula* and *Saxidomus*, according to Halstead (1978). Recently, Saitanu *et al.* (1991) found PSP toxins in the green mussel, *Perna viridis*, and the sand crab, *Asaphis violascens*, in the absence of any significant protistan bloom. The toxins killed mice with signs similar to PSP. On HPLC-fluorometric analysis, the mussel poison showed peaks with identical retention times to those of GTX 1–4, STX and neo-STX. The crab poisoning gave retention times identical to tetrodotoxin. According to the authors, these findings attest to the possible role of bacteria in PSP.

PSP has been reported following the ingestion of cultured purple clams *Coletellina diphos* in the Orient. The toxic fraction taken off the column was shown to contain gonyautoxin I, II, III and IV. Further purification gave a lethality level similar to that of gonyautoxin. The authors suggested that the chicken biventer cervicis nerve-muscle preparation is a suitable biological assay for field testing (Ho and Hwang, 1989).

Eating the ovaries of the Japanese callista, *Callista brevisiphonata*, has resulted in numerous cases of illness. Asano (1954) found that the ovaries of this species contained large amounts of choline but no histamine. Hashimoto (1979), thought that the choline was derived from cholinesters by the action cholinesterase, rather than from lecithin by the action of lecithinase. Cats fed the shellfish showed few signs, other than hypoactivity and some loss of coordination. Three of nine human volunteers who ate the ovaries developed urticaria and very mild symptoms.

Halstead (1965–70) states that venerupin poisoning was first reported in Nagai, Japan in 1889, following the ingestion of the oyster *Crassostrea gigas*. Of the 81 persons poisoned, 51 died. A second outbreak occurred in 1941, when five of six patients died, and from 1942 to 1950 there were 455 additional cases involving the eating of oysters and the short-necked clam *Tapes japonica* (Hashimoto, 1979). Akiba (1949) found that the toxin caused hemorrhage in the heart, lungs and viscera, with diffuse hemorrhage, necrosis and fatty degeneration of the liver. The toxin was named venerupin.

Akiba (1943) initiated a series of studies on the causative agent. Neutral methanol extracts of macerated tissues of *Tapes* and *Crassostrea* were subjected to absolute ethanol. The residue was then dissolved in methanol and again precipitated with ethanol. The toxic precipitate was found to be heat stable through the PH range 3.0–8.0, and resistant to boiling for three hours. No loss in toxicity was observed following storage in an ice box at pH 5.6 over a

20-month period. The toxin was found to be soluble in water, methanol, acetone and acetic acid. It was insoluble in benzene, ether and absolute alcohol. Certain heavy metals, picric acid and flavianic acid caused destruction of the toxin, It differed from mussel poison in that it could not be precipitated with reinecke acid. It was lethal to mice at the 0.25 mg/kg body weight level. Akiba suggested it may contain a double bond since it readily absorbed bromine. It was suggested that the toxin possessed a pyridine ring in the molecule (Akiba, 1949; Akiba and Hattori, 1949; Ishidate and Higiwara, 1954).

The earliest pharmacological studies on the sea hare were done by Flury (1915), using *Aplysia delipans* and *A. limacina*. He divided the secretions of *Aplysia* into three distinct types: a colorless secretion released from the surface of the animal; a terpene-like smelling, whitish, viscous material secreted from the opaline gland; and a purple substance from the purple gland. When applied to a frog's heart, the secretion from the body surface caused an arrhythmia, then slowing and finally cardiac standstill. The opaline secretion caused somewhat similar changes and when it was injected into several different marine animals and frogs, all developed muscular paralysis, and several died, tasting the opaline secretion. Flury found it bitter; it produced a slight burning sensation and irritated the oral mucosa, had no effect on the human skin, but when placed in the eye of a rabbit or dog it caused immediate irritation. Flury found the secretion neutral in pH. When the purple substance was injected into frogs, the animals suffered minor reflex changes and a transient paresis.

The acetone extracts of the digestive gland of *A. californica* were found to have an intraperitoneal LD_{50} of approximately 30 mg of sea hare tissue/kg mouse body weight (Winkler, 1961). Signs included increased respiration, blanching, drooping of the ears, increased salivation, muscle fasciculations, agonal signs, ataxia, prostration and death. The extract was also lethal when given orally at approximately 12 times the intraperitoneal dose. Using column chromatography, Winkler *et al.* (1962) obtained a partially purified poison, which they called aplysin, and found that in the dog the toxin had an immediate hypotensive effect. Recovery occurred when small doses were given. There was some initial arrhythmia followed by a slower but regular rate. In the isolated heart of the frog, aplysin caused standstill. The anterior cervical sympathetic ganglion of the cat was stimulated initially and then reversibly blocked. The frog rectus abdominis muscle responded by contraction and rat diaphragm-phrenic nerve neuromuscular junction was blocked; the block was antagonized by neostigmine. Rabbit intestinal muscle became spastic with temporary cessation of peristalsis, and the spasm was blocked by atropine. Washing of any of the preparations usually resulted in reversal of the changes.

In 1963, Yamamura and Hirata isolated two bromine-containing sesquiterpenes from whole Japanese sea hares, *A. kurodai*. They named these alpysin and aplysinol:

	R1	R_2
Aplysin	Br	CH_2
Aplysinol	Br	CH_2OH
Debromoaplysin	H	CH_3

They also obtained a debromo-derivative, debromoaplysin, and subsequently a third bromo compound, diterpene aplysin-20 was isolated (Matsuda and Tomue, 1967). In 1952, Ando had observed a sea hare feeding on the alga *Laurencia nipponica* and conducted several experiments which indicated that steam distillates of the alga were toxic to worms and the carp. Based on this observation, Irie *et al.* (1969) extracted aplysin, debromoaplysin and aplysinol from the red alga *Laurencia okamurai*, while Waraszkiewicz and Erickson (1974) obtained aplysin from *L. nidifica*. From these observations it was suggested that the bromo toxins originate in algae. Aplysiatoxin, debromoaplysiatoxin and 3-deoxydebromoaplysiatoxin are well known inflammatory agents.

Working with the digestive gland of *A. californica*, Faulkner *et al.* (1973) and Faulkner and Stallard (1973) isolated aplysin, debromoaplysin, laurenterol, johnstonol, pacifidiene and pacifidiene dichlorida. Since these compounds are found in *L. pacifica*, upon which this sea hare feeds, it has been suggested that these terpenes from the midgut may vary, depending on the algae on which the mollusc grazes (Hashimoto, 1979).

Two lethal extracts, one ether-soluble and other water-soluble in the digestive glands of the Hawaiian sea hares

Dolabella auriculasia, A. pulmonica, Stylocheilus longicauda and *Dolabrifera dolabriefa* were found by Watson (1973). Homogenized glands were extracted with acetone followed by partitioning between diethyl ether and distilled water. The ether-soluble fraction was chromatographed on a silicic acid column, while the water-soluble material was further extracted with 1-butanol. Both extracts appeared to be unaffected by brief exposure to temperature up to 90°C, as well as to recurrent freezing and thawing over periods exceeding two years. Both were most effective at pH 2.7, and inactive above pH 8.0. Neither raw glands nor their extracts were toxic orally to mice, but both ether- and water-soluble fractions produced effects following their intraperitoneal injection. Signs for injection of the ether-soluble toxin included irritability, viciousness and severe flaccid paralysis. The water-soluble toxin, in contrast, caused "convulsions" and respiratory distress.

Watson and Rayner (1973) observed that sublethal doses of the ether-soluble residue produced hypertension when injected intravenously into anesthetized rats, whereas the crude water-soluble residue produced a transient hypotension, bradycardia and apnea. Dose-effect curves for more purified extracts showed actions similar to those seen with crude extracts. The hypertension produced by the ether-soluble toxin was resistant to both alpha- and beta-adrenergic blocking agents. The hypotensive effect of the water-soluble extract could not be abolished by vagotomy or pretreatment with either atropine (25 mg/kg) or diphenhydramine (22 mg/kg). It was concluded that both extracts may have direct effects on the contractility of vascular smooth muscle which are not mediated by alpha-adrenergic or cholinergic mechanisms.

The ether-soluble toxin from *S. longicauda*, as described by Watson (1973), was further purified by Scheuer (1975) and Kato and Scheuer (1974, 1975). They found an oily mixture of aplysiatoxin and debromoaplysiatoxin following fractionation by silica column chromatography, cell filtration and thin layer chromatography (see top of next page).

In 1975, Blankenship *et al.* confirmed the observation of Winkler *et al.* (1962) relating to the presence of choline esters in the aqueous fraction of the digestive gland of *A. californica.* They identified both acetylcholine and urocanylcholine (murexine). The latter accounted for the

Aplysiatoxin   R = Br
Debromoaplysiatoxin   R = H

cholinesterase-resistant cholinomimetic activity of extracts of the gland. Kinnel *et al.* (1977) found that the "antifeedant" in *A. brasiliana* was the aromatic bromoallene, panacene. It was suggested, on the basis of studies by other workers, that the panacene is biosynthesized from a C_{15}, algal precursor. The provisional structure was:

Since the early studies of Bert (1867) and Lo Bianco (1888), an impressive number of substances has been isolated from or identified in the salivary glands of various cephalopods. Many of these substances were shown to have biological activities, although these activities were not always apparent in the physiopharmacological effects of the whole toxin, and some substances either did not have a significant biological activity or the then state of knowledge did not indicate what activity was present. Finally, the amounts of the various components from the salivary glands of cephalopods were subject to such variations that it was most difficult to determine whether or not a particularly toxic substance was present in a sufficient quantity to be deleterious to an envenomated animal. The importance of the synergistic effects of several of the toxic components, and of the autopharmacological responses, further complicated consideration of the chemistry and toxicology of these venoms. It now appears that many of the substances studied were actually normal constituents of the salivary glands and not necessarily venom constituents.

A few of the substances identified from extracts of *Octopus vulgaris* salivary glands prior to 1964 are shown in Table 22. The reader is referred to Russell (1965a) for the references to the identification of these substances. In the salivary glands of cephalopods, in general, tyramine, octopine, agmatine, adrenaline, noradrenaline, 5-hydroxytryptamine, L-ρ-hydroxyphenylethanolamine, histamine, dopamine, tryptophan, and certain of the 11-hydroxysteroids, polyphenols, phenolamines, indoleamines and guanidine bases have been identified.

The L-ρ-hydroxyphenylethanolamine was first described by Erspamer in 1940. It was found in extracts of the posterior salivary glands of *O. vulgaris* and associated with an adrenaline-like activity. It was thought to be the precursor of hydroxyoctopamine, or L-noradrenaline (Erspamer, 1952). Hartman *et al.* (1960) showed that the content of the posterior salivary glands of *O. apollyon* or *O. bimaculatus* decarboxylated L-3,4-dihydroxyphenyalanine (DOPA), DL-5-hydroxytryptophan, DL-*erythro*-3,4-dihydroxyphenyserine, DL-*erythro*-p-hydroxyphenylserine, DL-*m*-tyrosine, DL-*erythro*-m-hydroxyphenylserine, histidine, L-histidine, DL-*erythro*-phenylserine, 3,4-dihydroxyphenylserine, tyrosine and *m*-tyrosine. In general, the salivary glands of cephalopods were shown to contain few or no proteolytic enzymes, amylases or lipases; hyaluronidase was present in some glands.

Ghiretti (1959) purified a protein, cephalotoxin, from the posterior salivary glands of *Sepia officinalis*. He suggested this was the biologically active component of the toxin. It gave positive biuret and ninhydrin reactions, and had maximum ultraviolet absorption at 276–278 nm. Four bands migrating towards the cathode were observed on starch gel electrophoresis at pH 8.5. Further purification was obtained by absorption on calcium phosphate gel at neutral pH, and three bands were obtained on electrophoresis. Treatment with trypsin resulted in the complete loss of activity. The toxin contained no cholinesterase or aminoxidase activity. Analysis of cephalotoxin from the posterior salivary gland of *Octopus vulgaris* showed: protein 74.05 percent (N-determination), 64.25 percent (biuret reaction); carbohydrates, 4.71 percent and hexosamines, 5.80 percent (Ghiretti, 1960).

In 1949, Erspamer observed that the posterior salivary glands of *Eledone moschata* and *E. aldrovandi* contained a substance which when injected into mammals caused marked vasodilation, and produced hypotension and stimulation of certain extravascular smooth muscles. The substance was first called moschatin but was later renamed eledoisin. It was an endecapeptide having the following amino acid sequence:

Pyr-Pro-Ser-Lys-Asp (OH)$_1$-Ala-phe-Ileu-Gly-Leu-Met-NH$_2$

Subsequent studies showed that eledoisin was 50 times more potent than acetycholine, histamine or bradykinin in its ability to provoke hypotension in the dog. It produced an increase in the permeability of the peripheral vessels, stimulated the smooth muscles of the gastrointestinal tract and caused an increase, which was atropine-resistant in salivary secretions. It was easily distinguishable from kinins and substance P (Erspamer and Anastasi, 1962). In spite of its marked pharmacological activities, the role and significance of this substance in the salivary glands of *Eledone* is not clear. It was not found in the salivary glands of *O. vulgaris* or *O. macropus*, indicating that it is not a necessary component of cephalopod toxin. It appeared that eledoisin played some part in protein synthesis in the salivary gland (Table 23).

Reports of envenomation by Australian octopods, some of which were fatal (Mabbet, 1954; Flecker and Cotton, 1955; McMichael, 1963; Hopkins, 1964), stimulated renewed interest in the venom of cephalopods. Simon *et al.* (1964) prepared a saline extract of homogenized glands of *Hapalochlaena maculosa* and obtained a dialyzable, heat-stable product that resisted mild acid hydrolysis. This product was then studied on a number of pharmacological preparations. It was concluded that animals died in respiratory failure due to a phrenic nerve block and/or to deleterious changes at the neuromuscular junction. They also found that the extract produced bradycardia and hypotension, without remarkable changes in the electrocardiogram. A chromatographic study indicated the possible presence of octopamine and tryamine. Trethewie (1965) found that saline extracts of homogenized posterior glands of *H. maculosa* caused contraction of the guinea pig ileum, a decrease in the isolated cat heart rate, and depression of respirations,

Table 23. Activities of salivary glands of *Octopus vulgaris*

Activity	Salivary gland	
	Posterior	Anterior
Tyramine oxidase	X	
Tryptamine oxidase	X	
5-Hydroxytryptamine oxidase	X	
Proteolytic	X	
Hyaluronidase	X	X
Mucinolytic	X	X
Dopa decarboxylase	X	
Histamine oxidase	X	weak
Succinic dehydrogenase	X	X
Phosphatase	X	weak
Adenosine-triphosphatase	X	weak
Butyrylthiocholinesterase	X	0
Acetylthiocholinesterase	X	0
Acetylnaphtholesterase	X	X
Alpha naphtholase		weak

X Experimental evidence indicates activity.
0 Activity absent.
From Russell, 1965a. References in original table.

and hypotension and cardiac failure in the intact cat. In the rat phrenic nerve-diaphragm preparation the extracts caused almost immediate cessation of the indirectly stimulated contractions and severe impairment of the directly stimulated contractions. Although quantitative studies were not done, and the dose employed to be rather massive, it did appear that the toxin had a marked effect on neuromuscular transmission.

In 1969, Sutherland and Lane found that within minutes of placing a live *H. maculosa* on the back of rabbit a small bleeding puncture wound surrounded by a blanched area could be seen, the rabbit became restless, there was some exophthalmos and "one slight convulsion" with cessation of all muscular activity, other than cardiac. Cyanosis developed and death followed at 19 minutes. Thin layer chromatography of crude extracts of venom glands on silica gel gave nine spots, with the lethal spots having R_f values of 0.20 and 0.59. Dialysis removed all toxicity, leaving two high molecular weight components. Further studies with curtain electrophoresis and acrylamide gel electrophoresis again separated the two toxins but concentration studies were difficult. The authors concluded that there were two toxins, both having molecular weights below 1000 (and perhaps below 500), that a young specimen has sufficient venom in its posterior salivary glands to cause paralysis in 750 kg of rabbits, that the gland extracts have a high concentration of hyaluronidase, and that neostigmine does not reverse or reduce the toxic effects of the venom.

Freeman and Turner (1970) continued the earlier observations by studying an aqueous extract of gland tissue partially purified by filtration through Sephadex G-25. They obtained a lethal fraction that they termed maculotoxin, which they stated had a similar size and molecular configuration to saxitoxin. On the basis of lethality determinations they found a close similarity between their toxin and tetrodotoxin and concluded that maculotoxin appeared to resemble tetrodotoxin more closely than it did saxitoxin. Respiratory failure on injection of maculotoxin was characterized by loss of diaphragmatic activity which was evident before "failure of phrenic nerve volleys. . . . The vascularity of the diaphragm may permit of muscular axonal block before this is evident in the phrenic nerve."

Sutherland *et al.* (1970) prepared saline and water extracts of the homogenized whole glands of *H. maculosa* that produced paralysis and death at the 1 mg gland/2 kg rabbit level. Neither atropine and/nor neostigmine mixed with the toxin before administration altered the development of the paralysis. They demonstrated the production of antibodies to a non-toxic high molecular weight component but not for a toxic low molecular weight component, using similar techniques for both. They suggested a molecular weight below 540, which they felt accounted for the "total lack of antigenicity." As the writers have found with other venoms, the method of immunization for producing antibodies for different sized proteins varies and it is not always possible to obtain demonstrable antibodies for a small protein or a peptide using techniques best suited for large proteins. However, this might not be true for the low molecular weight protein described by these workers.

The following year it was found that maculotoxin blocked neuromuscular transmission in the isolated sciatic-sartorius nerve-muscle preparation of the toad by inhibiting the action potential in the motor nerve terminals, and that the toxin had no post-synaptic effect. It was suggested that the toxin may block action potentials by displacing sodium ions from negatively charged sites in the membrane (Dulhunty and. Gage, 1971). In 1972, Croft and Howden observed one major toxin, maculotoxin, and a minor one having similar chemical properties in the venom of *H. maculosa*. Their isolation technique for the posterior salivary glands is shown in Chart 3. The maculotoxin behaved as a cation of low molecular weight (< 700).

Although MacGinitie (1942) demonstrated the ability of the octopus to spew its salivary secretions and, indeed, R. Schweet and FER obtained venom samples with Professor MacGinitie using this technique in 1951, mechanical or electrical stimulation of the animal does not seem to have found much favor among scientists. This is unfortunate because, as is the case with other animals, when extracts of a venom gland are used instead of the venom, there is no doubt that normal constituents of the venom gland,

Chart 3. Isolation of maculotoxin. All fractions were tested for toxicity, for the presence of ninhydrin-positive material, and for sodium ions. The toxic fractions were evaporated *in vacuo* at 40°C (after Croft and Howden, 1972).

A

40 glands homogenized in 30 ml
CH_3OH for 2 min

↓

Filter through Whatman No. 541
paper

↓

Filtrate with 2 g of Mallinckrodt CC-7, 100–200 mesh silica gel and solvent evaporated *in vacuo* at 40°C

↓

Dry powder on silica gel column prepared by slurrying in 100 ml $CHCl_3$

↓

Column washed with 100 ml $CHCl_3$

↓

B

B

Column washed with 200 ml 1:3
$CH_3OH : CHCl_3$

↓

Maculotoxin eluted with 100 ml of 1:1 $Ch_3OH : CHCl_3$; 2 × 5 ml and 7 × 10 ml fractions collected when opaque zone reached bottom of the column; 2 mg of white powder obtained.

↓

Column washed with 100 ml CH_3OH

↓

A small amount of a second toxic component eluted with 100 ml 5 : 95 HOAc : CH_3OH

rather than components of the venom, are being identified. Many of the early works on marine toxins probably reflect this deficit. For this reason, the little note by Ballering *et al.* (1972) on a method for obtaining the saliva from the proboscis of *Octopus apollyon* is important.

Howden and Williams (1974) extracted tyramine, serotonin and histamine from the posterior salivary gland of *H. maculosa*. Their concentrations are shown in Table 24. The tyramine content is relatively low compared to that found in other species (Erspamer, 1948). The serotonin content is also comparatively low (Welsha and Moorhead, 1960), while the histamine content is consistent with that observed in other cephalopods.

In 1975, Jarvis *et al.* again demonstrated the chemical similarities between maculotoxin and tetrodotoxin. The two toxins could be partially purified by the same techniques and were not separable by ion exclusion chromatography. The recoveries of toxicity for both poisons followed the same pattern with respect to pH. Other chemical properties were identical or similar. The authors concluded that they found no chemical evidence distinguishing the two toxins from other. Crone *et al.* (1976) demonstrated that the methanol fraction contained a single toxin, maculotoxin, and again noted the similarities between maculotoxin and tetrodotoxin.

A second biologically active substance was isolated from the posterior gland of *H. maculosa* by Savage and Howden (1977). It was given the name hapalotoxin and was thought to have a molecular weight of less than 700. Its estimated lethal dose appeared to be about three times that of maculotoxin and the signs to death appeared similar, although the hapalotoxin was slower in acting. Some chemical properties of the two toxins are summarized in Table 25.

Table 24. Concentration of amines in the posterior salivary glands of *Hapalochlaena maculosa*

Sample number	Weight of glands (g)	Amount of amine present (µg/g of gland)		
		Histamine	Tyramine	Serotonin
1	0.522	41	104	46
2	0.663	50	65	36
3	0.710	25	92	31

(From Howden and Williams, 1974).

Table 25. Comparison of the properties of Hapalotoxin and Maculotoxin

Property		Hapalotoxin	Maculotoxin
Ambient stability		Unstable	Stable
Behavior on Sephadex G-10		Retarded	Retarded
u.v. spectrum		End absorption	λ_{max} 270
(nm)		near 220	$E^1_1 = 6.2$
TLC R_f system	A	0.20	0.65
	B	0.08	0.35
	C	0.30	0.50
TLC detection l_2		Positive	Positive
	KOH/MeOH	Positive	Positive
Ninhydrin		Negative	Positive

(From Savage and Howden, 1977).

In 1978, Scheumack *et al.* confirmed the previous observations on the similarities of maculotoxin to tetrodotoxin. Direct spectral and chromatographic comparisons showed the two toxins to be indistinguishable. This is of particular interest because here we have a poison (tetrodotoxin) which is also a venom (maculotoxin). In the former the presence of the poison is thought to be a product of metabolism. In the latter the venom is used to immobilize and perhaps kill the prey.

D.   Clinical Problem

Many cones have been implicated in injuries to man, including *C. geographus*, *C. aulicus*, *C. gloria-maris*, *C. marmoreus*, *C. textilis*, *C. tulipa*, *C. striatus*, *C. omaria*, *C. catus*, *C. obscurus*, *C. imperialis*, *C. pulicarius*, *C. quercinus*, *C. litteratus*, *C. lividus*, *C. sponsalis* and *C. californianus*. The first six named cones would seem the most dangerous as they are said to have the most highly developed venom apparatus (Halstead, 1978).

Cone stings often gives rise to immediate, sometimes intense, localized pain at the site of injury. Within five minutes the victim may note some numbness and ischemia about the wound, although in a case seen by one of us, the affected area was red and tender rather than ischemic. A tingling or numbing sensation may develop about the mouth, lips and tongue, and over the peripheral parts of the extremities, although this is not common. Other symptoms and signs during the first 30 minutes following the injury include: hypertoxicity, tremor, muscle fasciculation, nausea and vomiting, dizziness, increased lachrymation and salivation and weakness. The numbness about the wound may spread to involve a part of the extremity or injured part. In the more severe cases, respiratory distress with chest pain, difficulties in swallowing and phonation, marked dizziness, blurring of vision and an inability to focus, ataxia, and generalized pruritus have been reported. In fatal cases, "respiratory paralysis" precedes death (Russell, 1965a).

Poisoning following the ingestion of the whelk *Neptunea arthritica* appears to be a public health problem in parts of Japan. The clinical characteristics are headache, dizziness, nausea, vomiting, weakness, ataxia, photophobia, external ocular weakness, dryness of the mouth and, on occasion, urticaria.

While *Murex* species have been eaten in some parts of the world without ill effects (Dubois, 1903). Plumert (1902) cites a mass poisoning in the Gulf of Trieste in which 43 persons were poisoned after eating *M. brandaris*, and five persons died. There has been some question about this incident. However, Charnot (1945) states that ingestion of *M. brandaris* may produce gastroenteritis, pruritus, convulsions and death.

Latin and medieval writers from the time of Pliny considered the sea hare *Aplysia* to be very poisonous. Their beliefs have been reviewed by Johnston (1980). In spite of the reports of ancient writers that extracts of *Aplysia* were "frequently employed to dispatch their political enemies" (Halstead, 1965–70), there are no recent reports of deaths from the eating of sea hares. Tasting them produces a burning sensation in the mouth and slight irritation of the oral mucosa. Handling the animals is not likely to be dangerous. However, I would suspect that it would not be safe to rub one's eyes after handling these animals.

A number of poisonings occurred in Hokkaido, Japan, from the ingestion of *Callista brevisiphonata* in the early 1950s. These poisonings necessitated prohibiting the sale of the shellfish in the market place. The illness has a rapid onset, often occurring while the patient is still dining. It has been characterized as an "allergic-like" reaction, which is thought to be due to the presence of excessive choline in the ovaries. The most common findings are flushing, urticaria, and wheezing and gastrointestinal upset. It is self-limiting (Russell, 1971a,b).

Hashimoto (1979) notes a total of 542 cases of venerupin poisoning in Japan with 185 deaths. Fortunately, there have been no reported cases since 1950. It was observed following the eating of the oyster *Crassostrea gigas* or the asari *Tapes japonica*. The poisoning is characterized by a long incubation period (24–48 hours, sometimes longer), anorexia, halitosis, nausea, vomiting, gastric pain, constipation, headache and malaise (Togashi, 1943). These findings may be followed by increased nervousness, hematemesis, and bleeding from the mucous membranes of the nose, mouth and gums. In serious cases, jaundice may be present, and petechial hemorrhages and ecchymosis may appear over the chest, neck and arms. Leucocytosis, anemia, and a prolonged blood-clotting time are sometimes observed. The liver may be enlarged. In fatal poisonings, extreme excitation, delirium and coma occur. Venerupin shellfish poisoning is a public health problem in the Schizouoka and Kanagawa prefectures of Japan, where these shellfish are not eaten from January through April.

The more common types of shellfish poisoning are recognized as gastrointestinal, allergic or paralytic. Gastrointestinal shellfish poisoning is characterized by nausea, vomiting, abdominal pain, weakness and diarrhea. The onset of symptoms generally occurs 8 to 12 hours following ingestion of the offending mollusc. This type of intoxication is caused by bacterial pathogens, and is usually limited to gastrointestinal signs and symptoms. It rarely persists for more than 48 hours.

Allergic or erythematous shellfish poisoning is characterized by an allergic response, which may vary from one individual to another. The onset of symptoms and signs occurs 30 min. to six hours after ingestion of the mollusc to which the individual is sensitive. The usual presenting signs and symptoms are diffuse erythema, swelling, urticaria and pruritus involving the head and neck and then spreading to the body. Headache, flushing, epigastric distress and nausea are occasional complaints. In the more severe cases, generalized edema, severe pruritus, swelling of the tongue and throat, respiratory distress and vomiting sometimes occur. Death is rare but persons with known sensitivity to shellfish should avoid eating all molluscs. The sensitizing material appears more capable of provoking a serious autopharmacological response than most known sensitizing proteins.

Paralytic shellfish poisoning (PSP) is known variously as gonyaulax poisoning, paresthetic shellfish poisoning, mussel poisoning or mytilointoxication. Pathognomonic symptoms develop within the first 30 minutes following ingestion of the offending mollusc. Parasthesia, described as tingling, burning or numbness, is noted first about the mouth, lips and tongue; it then spreads over the face, scalp and neck, and to the fingertips and toes. Sensory perception and proprioception are affected to the point that the individual moves incoordinately and in a manner similar to that seen in another, more common form of intoxication. Ataxia, incoherent speech or aphonia are prominent signs of severe poisonings.

In some patients there is the complaint of dizziness, tightness of the throat and chest, and some pain on deep inspiration. Weakness, malaise, headache, increased salivation and perspiration, thirst, and nausea and vomiting may be present. The pulse is usually thready and rapid, the superficial reflexes are often absent and the deep reflexes may by hypoactive. If muscular weakness and respiratory distress grow progressively more severe during the first eight hours, death may ensue. If the victim survives the first 10 to 12 hours, the prognosis is good. Death is usually attributed to "respiratory paralysis" (Russell, 1965a, 1971a,b).

Among the cephalopods that have been implicated in bites on humans are *Hapalochaena (=Octopus) maculosa, Octopus australis, O. lunulatus, O. dofleini, O. vulgaris, O. apollyon, O. bimaculatus, O. macropus, O. rubescens, O. fitchi, O. flindersi, Ommastrephes sloani pacificus, Eledone moschata, E. aldrovandi* and *Sepia officinalis.*

The bites of most octopods results in a small puncture wound that appears to bleed more freely than one would expect from a similar non-envenomized traumatic wound (Fig. 46). Pain is minimal, and in the two cases seen by the authors it was described as no greater than that which would have been produced by a sharp pin. The area around the wound is first blanched but then becomes erythematous

Figure 46. Dislodging an octopus may be difficult, during which time the mollusc is more likely to bite.

and in severe envenomations it may become hemorrhagic. Tingling and numbness about the wound site are not uncommon complaints. Swelling is usually minimal immediately following the injury but may develop 6 to 12 hours later. Muscle fasciculations have been noted following *H. maculosa* bites (Sutherland and Lane, 1969). Localized pruritus sometimes occurs over the edematous area. "Lightheadedness" of several hours duration and weakness were reported in both cases observed by one of us; there were no other systemic symptoms or signs, and the wounds healed without complications (Russell, 1965a).

Flecker and Cotton (1955) report a patient bitten by *H. maculosa* who complained of dryness in the mouth and difficulty in breathing following the bite, but no localized or generalized pain. Subsequently, breathing became more labored, swallowing became difficult, and the patient began to vomit. Severe respiratory distress and cyanosis developed and the victim expired. The findings at autopsy were negative. Subsequently, Cleland and Southcott (1965) reviewed the literature on cephalopod bites and noted several unreported bites on humans. Additional cases have been described by Sutherland and Lane (1969) and Snow (1970).

Poisoning following the eating of cephalopods appears to be very rare. Kawabata *et al.* (1957) described 92 patients who became critically ill following the ingestion of *Ommastrephes sloani pacificus*. The patients developed nausea, vomiting, abdominal pain, diarrhea, fever, headache, chills and weakness. Three patients developed paralysis and three others had convulsions. In another series of poisonings there were 598 cases with five deaths. The species most likeiy involved were *Octopus vulgaris* Lamarck and *O. dofleini*. Intoxications took place between June to mid-September. In none of these cases was there any evidence of bacterial poisoning. Eating of the squid, in most instances raw or cooked, occurred within several hours of capture, while the octopods were eaten following boiling. No evidence of histamine or putrefactive amines were found in test samples.

REFERENCES

Abita, J.-P., Chicheportiche, R., Schweitz, H. *et al.* Effects of neurotoxins (veratridine, sea anemone toxin, tetrodotoxin) on transmitter accumulation and release by nerve terminals *in vitro. Biochemistry* 16, 1838, 1977.

Ackerman, D. Ueber die Extractstoffe von *Mytilus edulis.* I Mitteilung. *Z. Biol.* 74, 67, 1922.

Aguilar-Santos, G. and Doty, M.S. Chemical studies on three species of the marine algal species *Caulerpa.* In, *Drugs from the Sea.* Freudenthal, H.D. (ed.), p. 173, Marine Technical Society: Washington, DC, 1968.

Aiyar, R.G. Mortality of fish of the Malabar Coast in June 1935. *Curr. Sci.* 6, 688, 1936.

Akakawa, O., Onoue, Y., Miyazawa, K. *et al.* Tetrodoxin-bearing flatworms in southern Japan. *10th World Congress on Animal, Plant and Microbial Toxins,* Singapore. 1991.

Akiba, T. Study of poisons of *Venerupis semidecussata* and *Ostrea gigas. Nippon Iji Shimpo* (1087), 1077, 1943.

Akiba, T. Study on poisoning by the Short-necked clam and its poisonous substances. *Nissin Igaku* 36, 1, 1949.

Akiba, T. and Hattori, Y. Food poisoning caused by eating asari (*Venerupis semidecussata*) and oyster (*Ostrea gigas*) and studies on the toxic substance, venerupin. *Jpn. J. exp. Med.* 20, 271, 1949.

Alagarswami, K. and Narasimham, K.A. *Clam, Cockle and Oyster Resources of the Indian Coast.* CMFR Institute: Cochin, India, 1973.

Alam, J.M. and Qasim, R. Preliminary studies on biological and hazardous marine toxins from Karachi coast: (I) biochemical and biological properties of *Physalia* venom. Abstracts of 10th World Congress on Animal, Plant and Microbial Toxins, Singapore, #227, P. 280, 1991.

Alcala, A.C. and Halstead, B.W. Human fatality due to ingestion of the crab *Demania* sp. in the Philippines. *Clin. Tox.* 3, 609, 1970.

Alender, C.B. A biologically active substance from the spines of two diadematid sea urchins. In, *Animal Toxins.* Russell, F.E. and Saunders, P.R. (eds.), Pergamon: Oxford, 1967.

Alender, C.B., Feigan, G.A. and Tomita, J.T. Isolation and characterization of sea urchin toxin. *Toxicon* 3, 9, 1965.

Alender C.B. and Russell, F.E. Pharmacology. In, *Physiology of Echinodermata.* Boolootian, R.A. (ed.), Interscience: New York, 1966.

Alsen, C. and Reinberg, T. Characterization of the pharmacological and toxicological actions of two toxins isolated from the sea anemone *(Anemonia sulcata).* *Bull. Inst. Pasteur* 74, 117, 1976.

Anderson, D.M., White, A.W. and Baden, D.G. (eds.), *Toxic Dinoflagellates.* Elsevier: New York, 1985.

Anderson, D.M., Kulis, D.M., Sullivan, J.J. *et al.* Toxin composition variations in one isolate of the dinoflagellate *Alexandrium fundiense. Toxicon* 28, 885, 1990.

Ando, Y. Sea Hare Feeding behaviour. *Kagaku* 22, 87, 1952.

Anisimov, M.M., Fronert, E.B., Kuznetsova, T.A. *et al.* The toxic effect of triterpene glycosides from *Stichopus japonicus* Selenka on early embryogenesis of the sea urchin. *Toxicon* 11, 109, 1973.

Anisimov, M.M., Gafurov, N.N., Baranova, S.I. *et al.* [The influence of triterpenoid glycosides on activity of some enzymes of sub-cellular fractions of the rat liver and sea urchin embryos.] *Izv. Akad. Nauk. SSSR Ser. Biol.* 2, 301, 1978.

Anisimov M.M., Shcheglov, V.V., Stonik, V.A. *et al.* The toxic effect of cucumarioside-C from *Cucumaria fraudatrix* on early embryogenesis of the sea urchin. *Toxicon* 12, 327, 1974.

Anthoni, U., Bohlin, L., Larsen, C. *et al.* The toxin tetramine from the "edible" whelk *Neptunea antiqua. Toxicon* 27(7), 717, 1989.

Arndt, W. Über die Gifte der Plattwürmer. *Verhandl. Deut. zool. Ges.* 30, 135, 1925.

Arndt, W. Die Spongern als kryptotoxische Tiere. *Zool. Jahrb.* 45, 343, 1928.

Arndt, W. Polycladen and maricole Tricladen als Giftträger. *Mém. Estud. Mus. Zool. Univ. Coimbra* 148, 1, 1943.

Arndt, W. and Manteufel, P. Die Turbellarien als Träger von Giften. *A. Morphol. Oekol. Tiere* 3, 344, 1925.

Arnold, H.L., Jr., Grauer, I.H. and Chu, G.W.T.C. Seaweed dermatitis apparently caused by marine alga—clinical observations. *Proc. Hawaii Acad. Sci.* 34, 18, 1959.

Asano, M. Occurrence of choline in the shellfish, *Callista brevisiphonata* Carpenter. *Tohoku J. agric. Res.* 6, 147, 1954.

Asano, M. and Itoh M. Salivary poison of a marine gastropod, *Neptunea arthritica* Bernardi, and the seasonal variation of its toxicity. *Ann. N.Y. Acad. Sci.* 90, 647, 1960.

Ashok Kumar, B.A., Sarojini, R. and Nagabhushanam, R. Effects of holothurian *(Holothurea leucospilota)* toxin on the metabolism of the prawn *Caridina rajadhari.* In, *Bioactive Compounds from Marine Organisms.* Thompson, M.F., Sarojini, R. and Nagabhushanam, R. (eds.), p. 121. Oxford & IBH Publishing Co.: New Delhi, 1991.

Association of Official Analytical Chemists. Paralytic shellfish poisons biological method. *Official Methods of Analysis,* 12th ed., p. 319, Association of Official Analytical Chemists: Washington, D.C., 1975.

Atchinson, W.D., Luke, V.S., Narahashi, T. *et al.* Nerve membrane sodium channels as the target site of brevetoxin at neuromuscular junctions. *Br. J. Pharmacol.* 89, 731, 1986.

Attaway, D.H., and Ciereszko, L.S. In, *Proc. 2nd Int. Coral Reef Symp. 1.* Cameron, A.M. (ed.), Great Barrier Reef Committee, Brisbane, 1974.

Audebert, C. and Lamoureux, P. Eczéma professionel du marin pêcheur par contact de bryozoaires en Baie de Seine (premiers cas français 1975-1977). *Ann. Derm. Vénéréol.* 105, 187, 1978.

Autenreith, H.F. Uber das Gift der Fische. 387 pp. C.F. Ostander, Tubingen, 1833.

Bacq, Z.M. Les poisons de némertiens. *Bull. Cl. Sci., Acad. R. Belg.* (5)22, 1072, 1936.

Bacq, Z.M. L' "amphiporine" et la "némertine" poison des vers némertiens. *Arch. intern. Physiol.* 44, 190, 1937.

Baden, D.G. Brevetoxins: unique polyether dinoflagellate toxins. *F.A.S.E.B.* 3, 1807, 1989.

Baden, D.G. and Mende, T.J. Toxicity of two toxins from the Florida red tide marine dinoflagellate *Ptychodiscus breve. Toxicon* 20, 457, 1982.

Baden, D.G., Mende, T.J., Bikhazi, G. *et al.* Bronchoconstriction caused by Florida red tide toxins. *Toxicon* 20, 929, 1982.

Baden, D.G., Mende, T.J. and Brand, L.E. Cross reactivity in immunoassays directed against toxins isolated from *Ptychodiscus breve.* In, *Toxic Dinoflagellates.* Anderson, D.M., White, A.W. and Baden, D.G. (eds.) Elsevier: New York, 1985.

Baden, D.G., Mende, T.J., Waling, J. *et al.* Specific antibodies directed against toxins of *Ptychodiscus breve. Toxicon* 22, 783, 1984.

Baden, D.G. Mende, T.J., Szmant, A.M. *et al.* Brevetoxin binding: molecular pharmacology versus immunoassay. *Toxicon* 26, 97, 1988.

Baden, D.G., Mende, T.J. and Trainer, V.L. Derivatizated brevetoxins and their use as quantitative tools in detection. In, *Mycotoxins and Phycotoxins.* Natori, S., Hashimoto, K. and Veno, S. (eds.), p. 343, Elsevier: Amsterdam, 1989.

Bagnis, R. A case of coconut crab poisoning. *Clin. Tox.* 3, 585, 1970.

Bagnis, R., Legrand, A-M. and Inoue, A. Follow-up of a bloom of the toxic dinoflagellate *Gambierdiscus toxicus* on a fringing reef of Tahiti. *Toxic Marine Phytoplankton,* Graneli, E. (ed.), Elsevier Science Publishing Co.: Amsterdam, 1990.

Bahraoui, E.M., El Ayeb, M., Granier, C. *et al.* Specificity of antibodies to sea anemone toxin III, and immunogenicity of the pharmacological site of anemone and scorpion toxins. *Eur. J. Biochem.* 180, 55, 1989.

Bakus, G.J. Energetics and feeding in shallow-waters. *Int. Rev. gen. exp. Zool.* 4, 275, 1969.

Bakus, G.J. An ecological hypothesis for the evolution of toxicity in marine organisms. In, *Toxins of Animal and Plant Origin.* DeVries, A. and Kochva, E. (eds.), Gordon and Breach: London, 1971.

Bakus, G.J. Chemical defense mechanisms on The Great Barrier Reef, Australia. *Science* 211, 497, 1981.

Bakus, G.J. and Green, G. Toxicity in sponges and holothurians: a geographic pattern *Science* 185, 951, 1974.

Bakus, G.J. and Thun, M. Bioassays on the toxicity of Caribbean sponges. *Biol. Spongiaries* 291, 417, 1979.

Ballantine, D. and Abbott, B.C. Toxic marine flagellates, their occurrence and physiological effects on animals. *J. gen. Microbiol.* 16, 274, 1957.

Ballering, R.B, Jalving, M.A., Ventresca, D.E. *et al.* Octopus envenomation through a plastic bag via a salivary proboscis. *Toxicon* 10, 245, 1972.

Banner, A.H. A dermatitis-producing alga in Hawaii: preliminary report. *Hawaii Med.* 19, 35, 1959.

Banner, A.H. Ciguatera: a disease from coral reef fish. In: *Biology and Geology of Coral Reefs.* Jones, O.A. and Endean, R. (eds.), Academic Press: New York, 1976.

Banner, A.H. and Stephens, B.J. A note on the toxicity of the horseshoe crab in the Gulf of Thailand. *Nat. Hist. Bull. Siam Soc.* 21, 197, 1966.

Barnes, J.H. Observations on jellyfish stings in North Queensland. *Med. J. Aust.* 2, 993, 1960.

Barnes J.H. Extraction of cnidarian venom from living tentacle. In: *Animal Toxins.* Russell, F.E. and Saunders, P.R. (eds.), Pergamon: New York, 1967.

Baslow, M.H. *Marine Pharmacology.* 286 pp. Williams and Wilkins: Baltimore, MD 1969.

Bates, H.A. and Rapoport, H. A chemical assay for saxitoxin, the paralytic shellfish poison. *J. Agric. Food Chem.* 23, 237, 1975.

Batista, U., Macek, P. and Sedmax, B. The cytotoxic and cytolytic activity of equinatoxin II from the sea anemone *Actinia equina. Cell. Biol. Internat. Rpt.* 14, 1013, 1990.

Baxter, E.H. and Marr, A.G.M. Sea wasp *(Chironex fleckeri)* venom: lethal, haemolytic and dermonecrotic properties. *Toxicon* 7, 195, 1969.

Baxter, E.H., and Marr, A.G.M. Sea wasp toxoid: an immunizing agent against the venom of the box jelly fish, *Chironex fleckeri. Toxicon* 13, 423, 1975.

Baxter E.H., Walden, N.B. and Marr, A.G. Fatal intoxication of rabbits, sheep and monkeys by the venom of the sea wasp *(Chironex fleckeri). Toxicon* 10, 653, 1972.

Becker, S., Atherton, E. and Gorgon, R.D. Synthesis and characterization of μ-conotoxin IIIa. *Eur. J. Biochem.* 185, 79, 1989.

Bender, J.A., DeRiemer, K., Roberts, T.E. *et al.* Choline esters in the marine gastropods *Nucella emarginata and Acanthina spirata*: a new choline ester tentatively identified as N-methylmurexine. *Comp. gen. Pharmac.* 5, 191, 1974.

Bengston, K., Nichols, M.M., Schaadig, V. *et al.* Sudden death in a child following jellyfish envenomation by *Chiropsalmus quadrumanus. JAMA,* 226(10), 1404, 1991.

Berdyshev, E.V., Vaskovsky, V.E., Pavlenko, A.F. *et al.* Possible role of platelet activation factor (PAF) in Medusae toxicity. 10th World Congress on animal, plant and Microbial Toxins, Singapore. Abst. 196, 248, 1991.

Beress, L., Beress, R. and Wunderer, G. Purification of three polypeptides with neuro- and cardiotoxic activity from the sea anemone *Anemonia sulcata. Toxicon* 13, 359, 1975.

Bergmann, W. Sterols: their structures and distribution. In, *Comprehensive Biochemistry.* Florkin, M. and Mason, H.S. (eds.), Academic Press: New York and London, 1962.

Bergmann, W. and Feeney, R.J. Contributions to the study of marine products XVI. Chondrillasterol. *J. org. Chem.* 13, 738, 1951.

Bergmann, W. and Domisky, I.I. Sterols of some invertebrates. *Ann. New York. Acad. Sci.* 90, 906, 1960.

Bergmann, F., Parnas, I. and Reich, K. The action of the toxin of *Prymnesium parvum* Carter on the guinea-pig ileum. *Brit. J. Pharmacol.* 22, 47, 1964.

Bernheimer, A.W. and Avigad, L.S. Properties of a toxin from the sea anemone *Stoichactis helianthus*, including specific binding to sphingomyelin. *Proc. Natn. Acad. Sci. USA* 73, H67, 1976.

Bernheimer, A.W. and Avigad, L.S. A cholesterol-inhibitable cytolytic protein from the sea anemone *Metridium senile. Biochim. Biophys. Acta* 541, 96, 1978.

Bernheimer, A.W. and Lai, C.Y. Properties of a cytolytic toxin from the sea anemone, *Stoichactis kenti. Toxicon* 23, 791, 1985.

Berquist, P.R. *Sponges.* Univ. California Press: Berkeley, California, 1978.

Berquist, P.R. and Hartman, W.D. Free amino acid patterns and the classification of Demospongia. *Mar. Biol.* 3, 347, 1969.

Berquist, P.R. and Hogg, J.J. Free amino acid patterns in demospongiae: a biological approach to sponge classification. *Cah. Biol. Mar* 10, 205, 1969.

Bert, P. Mémoire sur la physiologie de la seiche. *Mem. Soc. Sci. Phys. Nat. Bordeaux* 5, 115, 1867.

Bheemasankara Rao, Ch., Venkata Rao, D., Kalidindi, R.S.H.S.N. *et al.* Furranoterpenes from *Phyllospongia dendyi* Lendenfeld of the Bay of Bengal. In, *Bioactive Compounds from Marine Organisms.* Thompson, M.F., Sarojini, R. and Nagabushanam, R. (eds.), Oxford & IBH Publishing Co.: New Delhi, 1991.

Bhimachar, B.S. and George, P.C. Abrupt set-backs in the fisheries of the Malabar and Karwar Coasts and red water phenomena as their probable cause. *Proc. Indian Acad. Sci.* 31B, 339, 1950.

Bisset, M.G. Hunting poisons in the North Pacific region. *Lloydia* 39, 87, 1976.

Blankenship, J.E., Langlais, P.J. and Kittredge, J.S. Identification of a cholinomimetic compound in the digestive gland of *Aplysia californica. Comp. Biochem. Physiol.* 51C, 129, 1975.

Blanquet, R.S. A toxic protein from the nematocysts of the scyphozoan medusa *Chrysaora quinquecirrha. Toxicon* 10, 103, 1972.

Blumenthal, K.M. and Kem, W.R. Structure-function relationships in *Cerebratulus* toxin B-IV. In, *Natural Toxins.* Eaker, D. and Wadström. T. (eds.), Pergamon Press: Oxford, 1980.

Blunden, G., Barwell, C.J., Fidgen, K.J. *et al.* A survey of British marine algae for anti-influenza virus activity. *Botanica Marina* 24(5), 267, 1981.

Bolton, B.L., Bergner, A.D., O'Neill, J.J. *et al.* Effect of a shell-fish poison on end-plate potentials. *Bull. Johns Hopkins Hosp.* 150, 233, 1959.

Bond, R.M. and Medcof, J.C. Epidemic shellfish poisoning in New Brunswick. In, *Conference on Shellfish Toxicology.* U.S. Publ. Health Svce. (Spec. Publ. 96): 1957.

Bonnevie, P. Fisherman's "Dogger Bank itch" an allergic contact-eczema due to the coralline *Alcyonidium hirsutum. Acta. Allergol.* 1, 40, 1948.

Bordner, J., Thiessen, W.E., Bates, H.A. and Rapoport, H. The structure of a crystalline derivative of saxitoxin. *J. Am. Chem. Soc.* 97, 6008, 1975.

Bottard, A. *Les Poissons Venimeux.* 198 pp. Doin: Paris, 1889.

Boyer, G.L., Shantz, E.J. and Schnoes, H.K. Characterization of 11-hydroxysaxitoxin sulphate, a major toxin in scallops exposed to blooms of the poisonous dinoflagellate *Gonyaulax tamarensis. Chem. Commun.* 889, 1978.

Braekman, J.C. and Daloze, D. Chemical defence in sponges. *Pure appl. Chem.* 53, 357, 1986.

Braekman, J.C., Daloze, D., de Abreu, P.M. *et al.* A novel type of bidquinosolizidine alkaloid from the sponge *Petrosia iserriata. Tetrahed. Lett.* 23, 4277, 1982.

Brand, D.D., Blanquet, R.S. and Phelan, M.A. Collagenaceous thiol-containing properties of cnidarian nematocysts. A comparison of the chemistry and protein distribution patterns in two types of Cnidae. *Comp. Biochem. Physiol. B. Comp. Biochem.* 106 B, 115, 1993.

Brieger, L. Zur Kenntniss des Tetanin und des Mytilotoxin. *Virchows Arch.* 112, 549, 1888.

Brongersma-Sanders, M. Mass mortality in the sea. *Geol. Soc., Amer. Memoir* 67(1), 941, 1957.

Bruni, J.E., Bose, R., Pinsky, C. *et al.* Circumventricular organ origin of domoic acid-induced neuropathology and toxicology. *Brain Res. Bull.* 26, 419, 1991.

Buckley, E.E. and Porges, N. *Venoms.* AAAS: Washington, D.C., 1956.

Burke, J.M., Marichisotto, J., McLaughlin, J.J.A. *et al.* Analysis of the toxin produced by *Gonyaulax catenella* in axenic culture. *Ann. Rev. N.Y. Acad. Sci.* 90, 837, 1960.

Burkholder, P.R. and Sharma, G.M. Antimicrobial agents from the sea. *Lloydia* 32, 466, 1969.

Burnett, J.W. and Calton, G.J. The chemistry and toxicology of some venomous pelagic coelenterates. *Toxicon* 15, 177, 1977.

Burnett, J.W., Stone, J.H., Pierce, L.H. *et al.* A physical and chemical study of sea nettle nematocysts and their toxin. *J. Invest. Derm.* 51, 330, 1968.

Burnett, J.W. The use of verapamil to treat box-jellyfish stings. *Med. J. Austr.* 153, 363, 1990.

Burnett, J.W. and Cable, W.D. A fatal jellyfish envenomation by the Portuguese Man-O' War. *Toxicon* 27(7), 823, 1989.

Burreson, B.J., Christophersen, C. and Scheuer, P.J. Cooccurrence of a terpenoid isocyanideformamide pair in the marine sponge *Halichondria* sp., *J. Am. Chem. Soc.* 97, 201, 1975.

Burreson, B.J., Moore, R.E. and Roller, P.P. Volatile halogen compounds in the alga *Asparagopsis toxiformis* (Rhodophyta). *J. Agric. Food Chem.* 24, 856, 1976.

Burton, M. Porifera. In, *The Encyclopedia of the Biological Sciences.* Gray, P. (ed.), p. 825. Reinhold: New York, 1961.

Calton, G.J. and Burnett, J.W. Partial purification and characterization of the alkaline protease of sea nettle (*Chrysaora quinquecirrha*) nemato-cyst venom. *Comp. Biochem. Physiol.* 74C(2), 361, 1983.

Cardellina, J.H.H., Marner, F.J. and Moore, R.E. Seaweed dermatitis: structure of lyngbyatoxin A. *Science* 204, 193, 1979.

Cariello, L. and D' Aniello, A. Isolation and characterization of four toxic protein fractions from the sea anemone *Anemonia sulcata. Toxicon* 13, 353, 1975.

Cariello, L. and Zanetti, L. Suberitine, the toxic protein from the marine sponge *Suberites domuncula. Comp. Biochem. Physiol.* 64C, 15, 1979.

Cariello, L., Salvato, B. and Jori, G. Partial characterization of suberitine, the neurotoxic protein purified from *Suberites domuncula. Comp. Biochem. Physiol.* 76B, 337, 1980.

Cariello, L., deSantis, A., Fiore, F. *et al.* Callitoxin, a neurotoxic peptide from the sea aneamone *Calliactis parasitica*: amino acid sequence and electrophysical properties. *Biochemistry* 28, 2484, 1989.

Carlé, J.S. and Christophersen, C. Dogger Bank Itch. The allergen is (2-hydroxyethyl) dimethyl-sulfoxonium ion. *J. Am. Chem. Soc.* 102, 5107, 1980.

Carlé, J.S., Thybo, H. and Christophersen, C. Dogger Bank itch (3). Isolation, structure determination and synthesis of a hapten. *Contact Derm.* 8, 43, 1982.

Carrilo N.C. *Boletin de Soc. Geogr. Lima* 2, 16, 1892.

Carter, N. New or interesting algae from brackish water. *Arch. Protistemis* 90, 1, 1938.

Catterall, W.A. The voltage sensitive sodium channel: a receptor for multiple neurotoxins. In, *Toxic Dinoflagellates*. Anderson, M., White, A.W. and Baden, D.G. (eds.), p. 329, Elsevier: New York, 1985.

Chanley, J.D., Mezzetti, T. and Sobotka, H.E. The holothurinogenins. *Tetrahedron* 22, 1857, 1966.

Chanley, J.D., Perlstein, J. Nigrelli, R.F. *et al.* Further studies on the structure of holothurin. *Ann. N.Y. Acad. Sci.* 90, 202, 1960.

Charnot, A. La toxicologie au Maroc. *Mém. Soc. Sci. nat. Maroc* 47, 86, 1945.

Chevallier, A. and Duchesne, E.A. Mémoire sur les empoisonnements par les huitres. les moules, les crabes, et par certains poissons de mer et de rivière. *Ann. Hyg. Pub.* 46, 108, 1851.

Cheymol, J. De dos substancias biomarinas inhaiboras neuromusculares: tetrodotoxina y saxitoxina. *Archos Fac. Med. Madrid* 8, 151, 1965.

Chia, D.G.B, Lau, C.O., Tan, C.H. *et al.* Localization of toxin in the poisonous marine crab, *Lophozozymus pictor. Abstracts of 10th World Congress on Animal, Plant and Microbial Toxins*. Singapore. Abstract 202, p. 254, 1991.

Chiszar, D., Smith, H.M. and Hoge, A.R. Post-strike trailing behavior in rattlesnakes. *Mem. Inst. Butantan* 46, 195, 1982.

Cimino, G., De Rosa, S., De Stefano, S. *et al.* Marine natural products: new results from Mediterranean invertebrates. *Pure Appl. Chem.* 58, 375, 1986.

Cimino, G., DeStefano, S. Fenical, W. *et al.* Zonaroic acid from the brown seaweed *Dictyopteris undulata (= zonarioides). Experientia* 31, 1250, 1975.

Clark, H.L. The echinoderm fauna of Torres Strait: its composition and its origin: Carnegie Inst. Washington, Publ. 214, 155, 1921.

Clasbrummell, B., Oswald, H. and Illes, P. Inhibition of noradrenaline release by w-conotoxin GVIA in the rat tail artery. *Br. J. Pharmacol.* 96, 101, 1989.

Cleland, J.B. Injuries and diseases of man in Australia attributed to animals (except insects). *J. trop. Med. Hyg.* 16, 25, 1913.

Cleland, J.B. Injuries and diseases in Australia attributable to animals. *Med. J. Aust.* (2), 313, 1942.

Cleland, J.B. *Injuries and Diseases of Man in Australia Attributable to Animals, Except Those Due to Snakes and Insects*. 6th Rept., Govt. Bur. Microbiol. Dept. Publ. Hlth.: N.S.W. Australia, 1961.

Cleland, J.B. and Southcott, R.V. *Injuries to Man from Marine Invertebrates in the Australian Region*. Commonwealth of Australia: Canberra, 1965.

Clemons, G.P., Pham, D.V. and Pinion, J.P. Insecticidal activity of *Gonyaulax* (Dinophyceae) cell powders and saxitoxin to the German cockroach. *J. Phycol.* 16, 305, 1986a.

Clemons, G.P., Pinion, J.P., Bass, E. *et al.* A hemolytic principle associated with the red-tide dinoflagellate *Gonyaulax monilata. Toxicon* 18, 323, 1980b.

Cocamese, S., Azzolena, G., Furnari, G. *et al.* Antimicrobial and antiviral activities of some marine algae from eastern Sicily. *Botanica Marina* 24(7): 365, 1981.

Cohen, S.S. Sponges, cancer chemotherapy and cellular aging. *Perspect. Biol. Med.* 6, 215, 1963.

Cohen, A.S. and Olek, A.J. An extract of lionfish (*Pterois voiitans*) spine tissue contains acetylocholine and a toxin that affects neuromuscular transmission. *Toxicon* 27, 1367, 1989.

Collins, R.C. Neurotoxins and the selective vulnerability of brain. In, *Neurotoxins and their Pharmacological Implications.* Jenner, P. (ed.), p. 1, Raven Press: New York, 1987.

Collins, S.P., Comis, A., Marshall, M. *et al.* Monoclonal antibodies neutralizing the haemolytic activity of box jellyfish (*Chironex fleckeri*) tentacle extracts. *Comp. Biochem. Physiol. B.* 106(1) 67, 1993.

Comis, A., Hartwick, R.F., Howden, M.E.H. Stabilization of lethal and hemolytic activities of box jellyfish (*Chironex fleckeri*) venom. *Toxicon* 27, 439, 1989.

Connell, C.H. and Cross, J.B. Mass mortality of fish associated with protozoan *Gonyaulax* in the Gulf of Mexico. *Science* 112, 259, 1950.

Cooper, H.S. What Bêche-de-mer is, how it is caught and what is done with it. In, *Coral Lands.* Copper, H.S., London, 1880.

Cormier, S.M. and Hessinger, D.A. Cnidocil apparatus: sensory receptor of *Physalia* nematocytes. *J. Ultrastruct. Res.* 72, 13, 1980a.

Cormier, S.M. and Hessinger, D.A. Cellular basis for the tentacle adherence in the Portuguese man-of-war (*Physalia physalis*). *Tissue and Cell* 12, 713, 1980b.

Creasia, D.A. and Neally, M.L. Acute isolation of saxitoxin in the mouse. *Toxicon* 27, 39, 1989.

Croft, J.A. and Howden, M.E.H. Chemistry of maculotoxin: a potent neurotoxin isolated from *Hapalochlaena maculosa. Toxicon* 10, 645, 1972.

Croft, J.A. and Howden, M.E.H. Isolation and partial characterization of a steroidal saponin from the starfish *Patiriella calcar. Comp. Biochem. Physiol.* 48B 535, 1974.

Crone, H.D., Leake, B., Jarvis, M.W. *et al.* On the nature of "Maculotoxin," a toxin from the blue-ringed octopus (*Hapalochlaena maculosa*). *Toxicon* 14, 423, 1976.

Crone, H.D. and Keen, T.E.N. Chromatographic properties of the hemolysin from the cnidarian *Chironex fleckeri. Toxicon* 7, 79, 1969.

Cruz, L.J. Conotoxins: biochemical probes for ion channels and receptors. *Toxicon* 27, 23, 1989.

Culver, P. and Jacobs, R.S. Lophotoxin: a neuromuscular acting toxin from the sea whip (*Lophogorgia rigida*). *Toxicon* 19, 825, 1981.

Dafni, Z. and Gilberman, E. Nature of initial damage to Erlich ascites cells caused by *Prymnesium parvum* toxin. *Biochim. Biophys. Acta* 255, 380, 1972.

Dale, B. and Yentsch, C.M. Red tide and paralytic shellfish poisoning. *Oceanus* 21, 41, 1978.

Das, N.P., Lim, H.S. and Teh, Y.F. histamine and histamine-like substances in the marine sponge, *Suberites inconstans. Comp. gen. Pharmac.* 2, 473, 1971.

De Groof, R.C. and Narahashi, T. The effects of holothurin A on the resting membrance potential and conductance of squid axon. *Eur. J. Pharmacol.* 36, 337, 1974.

De Laubenfels, M.W. The marine and freshwater sponges of California. *Proc. U.S. Natn. Mus.* 81, 1, 1932.

De Luca, A., Li, C.G., Rand, M.J. *et al.* Effects of ω-conotoxin GVIA on autonomic neuroeffector transmission in various tissues. *Br. J. Pharmacol.* 101, 437, 1990.

De Mendiola, B.R. Red tide along the Peruvian coast. In, *Toxic Dinoflagellate Blooms.* Taylor, D.L. and Seliger, H.H. (eds.), Elsevier: North-Holland, New York, 1979.

Dettbarn, W.D., Higman, H.B., Bartels, E. *et al.* Effects of marine toxins on electrical activity and K⁺ efflux on excitable membranes. *Biochim. Biophys. Acta* 94, 472, 1965.

Devlin, J.P. Isolation and partial purification of hemolytic toxin from sea anemone *Stoichactis helianthus J. Pharm. Sci.* 63, 1478, 1974.

Dickey, R.W., Bobzin, S.C., Faulkner, D.J. *et al.* Identification of okadaic acid from a Caribbean dinoflagellate, *Prorocentrum concavum. Toxicon* 28(4), 371, 1990.

Dimanlig, M.M.V. and Taylor, F.J.R. Extracellular bacteria and toxin production in *Protogonyaulax* species. In, *Toxic Dinoflagellates,* Anderson, D.M., White, A.W. and Baden, D.G. (eds.), p. 103, Elsevier: New York, 1985.

Dodge, J.D. Fine structure of the pyrrhophyta. *Bot. Rev.* 37, 481, 1971.

Dodge, J.D. *The Atlas of Dinoflagellates.* Farrand Press: London, 1985.

Doig, M.T., III and Martin, D.F. Physical and chemical stability of ichthyotoxins produced by *Gymnodinium breve. Environ. Lett.* 3, 279, 1973.

Doyle, J.W., Kem, W.R. and Villalonga, F.A. Interfacial activity of an ion channel-generating protein cytolysin from the sea anemone *Stichodactyla helianthus. Toxicon* 27(4), 465, 1989.

Driscoll, P.C., Clore, G.M., Beress, L. *et al.* A proton nuclear magnetic resonance study of the antihypertensive and antiviral protein BDS-I from the sea anemone *Anemonia sulcata:* sequential and sterospecific resonance assignment and secondary structure. *Biochemistry* 28, 2178, 1989.

Droop, M. A note on the isolation of a small marine alga and flagellates in pure culture. *J. Mar. biol. Ass. U.K.* 33, 511, 1954.

Drury, J.K., Noonan, J.D., Pollock, J.G. *et al.* Jellyfish sting with serious hand complications. *Injury* 12, 66, 1980.

Dubois, R. Sur le venin de la glande à pourpre des *Murex. C.R. Soc. Biol.* 55, 81, 1903.

Dubos, M., Nguyen, T.L., Lamoureux, P. *et al. Alcyonidium gelatinosum* (L.) (Bryozoaire) et réactions cutanées d' hypersensibilité, résultats preliminaires d' une étude expérimentale. *Bull. Soc. Path. Exot.* 70, 82, 1977.

Dulhunty, A. and Gage, P.W. Selective effects of an octopus toxin on action potential. *J. Physiol.* 218, 433, 1971.

Durand, M. and Puisen-Dao, S. Physiological and ultrastructural features of the toxic dinoflagellate *Gambierdiscus toxicus* in culture. In, *Toxic Dinoflagellates.* Anderson, D.M., White, A.W. and Baden, D.G. (eds.), p. 61, Elsevier: New York, 1985.

Durand, M. and Barkaloff, C. Pigment composition and chloroplast organization of *Gambierdiscus toxicus* A & F (Dinophyceae). *Phycol.* 24(2), 217, 1985.

Eckert, G.J. Absence of toxin-producing parapodial glands in amphinomid polychetes (fireworms). *Toxicon* 23(2), 350, 1985.

Elyakov, G.B. and Peretolchin, N.V. Cucumarioside-C—a new triterpenic glycoside from a holothurian—*Cucumaria fraudatrix. Khim. Prir. Soedin* 6, 637, 1970.

Endean, R. The cuvierian tubules of *Holothuria leucospilota. Q.J. microsc. Sci.* 98, 455, 1957.

Endean, R. A new species of venomous echinoid from Queensland waters. *Mem. Queensland Mus.* 14, 95, 1964.

Endean, R. and Duchemin, C. The venom apparatus of *Conus magus. Toxicon* 4, 275, 1967.

Endean, R., Duchemin, C., McColm, D. *et al.* A study of the biological activity of toxic material derived from nematocysts of the cubomedusan *Chironex fleckeri. Toxicon* 6, 179, 1969.

Endean, R. and Henderson, L. Further studies of toxic material from nematocysts of the cubomedusan *Chironex fleckeri* Southcott. *Toxicon* 7, 303, 1969.

Endean, R. and Izatt, J. Pharmacological study of the venom of the gastropod *Conus magus. Toxicon* 3, 81, 1965.

Endean, R., Izatt, J. and McColm, D. The venom of the piscivorous gastropod *Conus striatus.* In, *Animal Toxins.* Russell, F.E. and Saunders, P.R. (eds.), Pergamon Press: Oxford, 1967.

Endean, R. and Noble, M. Toxic material from the tentacles of the cubomedusan *Chironex fleckeri. Toxicon* 9, 255, 1971.

Endean, R. and Rudkin, C. Studies of the venoms of some Conidae. *Toxicon* 1, 49, 1963.

Endo, M., Nakagawa, M., Hamamoto Y. *et al.* Pharmacologically active substance from Southern Pacific marine invertebrates. *Pure appl. Chem.* 58, 387, 1986.

Erman, A. and Neéman, I. Inhibition of phosphofructokinase by the toxic cembranolide sarcophine insolated from the soft bodied coral *Sarcophyton glaucum. Toxicon* 15, 207, 1977.

Erspamer, V. Active substances in the posterior salivary glands of Octopoda. I. Enteramine-like substances. *Acta pharmacol. Toxicol.* 4, 213 1948.

Erspamer, V. Identification of octopamine as l-p-hydroxyphenylethanolamine. *Nature* 169, 375, 1952.

Erspamer, V. and Anastasi, A. Structure and pharmacological actions of eledoisin, the active endecapeptide of the posterior salivary glands of *Eledone. Experientia* 18, 58, 1962.

Erspamer, V. and Asero, B. Identification of enteramine, the specific hormone of the entero-chromaffin cell system, as 5-hydroxytryptamine. *Nature* 169, 300, 1952.

Erspamer, V. and Benati, O. Identification of murexine as b-(imidazoyl- [4])-acryl-choline. *Science* 117, 161, 1953.

Erspamer, V. and Dordoni, F. Ricerche chimiche farmacologiche suglie stratti di ghiandola ipobranchiale di *Murex trunculus, Murex brandarise Tritonalia erinacea. III. Arch. intern. Pharacodyn.* 74, 263, 1947.

Evans, M.H. Block of sensory nerve conduction in the cat by mussel poison and tetrodotoxin. In, *Animal Toxins.* Russell, F.E. and Saunders, P.R. (eds.), p. 47 Pergamon: Oxford, 1967.

Evans, M.H. Topical applications of saxitoxin and tetrodotoxin to peripheral nerves and spinal roots in the cat. *Toxicon* 5, 289, 1968.

Evans, M.H. Saxitoxin and related poisons: their actions on man and other mammals. In, *Proceedings of the First International Conference on Toxic Dinoflagellate Blooms*. LoCicero, V.R. (ed.), Mass. Sci. Tech. Found: Wakefield, Mass., 1975.

Exton, D.R., Fenner, P.J. and Williamson, J.A. Cold packs: effective topical analgesia in the treatment of painful stings by *Physalia* and other jellyfish. *Med. J. Aust.* 151, 625, 1989.

Fänge, R. The salivary gland of *Neptunea antiqua* (Gastropoda). *Acta Zool.* 39, 39, 1960.

Farber, L. and Leske, P. Studies on the toxicity of *Rhodactis howesii* (Matamaw). In, *Venomous and Poisonous Animals and Noxious Plants of the Pacific Region*. Keegan, H.L. and Macfarland, W.V. (Eds.), Pergamon Press: Oxford, 1963.

Faulkner. D.J. and Stallard, M.O. 7-chloro-3, 7-dimethy-1, 4, 6-tribromo-1-octen-3-ol, a novel monoterpene alcohol from *Aplysia californica. Tetrahedron Lett.* 14, 1171, 1973.

Faulkner, D.J., Stallard, M.O., Fayos, J. *et al.* A novel monoterpene from the sea hare, *Aplysia californica. J. Am. Chem. Soc.* 95, 3413, 1973.

Feder, H.M. Gastropod defensive responses and their effectiveness in reducing predation by starfishes. *Ecology* 44, 505, 1963.

Feder, H.M. The food of the starfish *Pisaster ochraceus* along the California coast. *Ecology* 40, 721, 1959.

Feder, H.M. and Arvidsson, J. Studies on a sea-star (*Marthasterias glacialis*) extract responsible for avoidance reactions in the gastropod (*Buccinum undatum*). *Arkiv Zool.* 19, 369, 1967.

Feigen, G.A., Sanz, E. and Alender, C.B. Studies on the mode of action of sea urchin toxin-I. Conditions affecting release of histamine and other agents from isolated toxins. *Toxicon* 4, 161, 1966.

Feigen, G.A., Sanz, E., Tomita, J. et al. Studies on the mode of action of sea urchin toxin-II. Enzymatic and immunological behavior. *Toxicon* 6, 17, 1968.

Feigen, G.A. Hadji, L. and Cushing, J.E. Modes of action and identities of protein constituents in sea urchin toxin. In, *Bioactive Compounds from the Sea*. Humm, H.J. and Lane, C.E. (eds.), Marcel Dekker: New York, 1974.

Fenical, W., Okuda, R.K., Bandurraga, M.M. *et al.* Lophotoxin: a novel neuromuscular toxin from Pacific sea whips of the genus *Lophogorgia*. *Science* 212, 1512, 1981.

Ferlan, I. Biochemical and biological characteristics of equinatoxin isolated from *Actinia equina*. Dissertation, Univ. Ljubljana: Yugoslavia, 1977.

Ferlan, I. and Lebez, D. Preliminary studies on the structure of equinatoxin. *Bull Inst. Pasteur* 74, 121, 1976.

Fenner, P.J., Williamson, J.A. and Blenkin, J.A. Successful use of *Chironex* antivenom by members of the Queensland Ambulance Transport Brigade. *Med. J. Aust.* 151, 798, 1989.

Ferlan, I. and Lebez, D. Equinatoxin, a lethal protein from *Actinia equina*-I. Purification and characterization. *Toxicon* 12, 57, 1974.

Fingerman, M., Forester, R.H. and Stover, J.H., Jr. Action of shellfish poison on peripheral nerve and skeletal muscle. *Proc. Soc., exp. Biol. Med.* 84, 643, 1953.

Fish, C.J. and Cobb, M.C. Noxious marine animals of the central and western Pacific Ocean. *U.S. Fish. Wildl. Res. Rep.* 36, 1954.

Fusetani, N. and Hashimoto, K. Diethyl peroxides probably responsible for *moyuku* poisoning. *Bull. Jpn. Soc. Scient. Fish.* 47, 1059, 1981.

Fusetani, N., Hashimoto, K., Mizukami, I. *et al.* Lethality in mice of the coconut crab *Birgus latro. Toxicon* 18, 694, 1980.

Fusetani, N., Ozawa, C. and Hashimoto, Y. Fatty acids as ichthyotoxic constituents of a green alga *Chaetomorpha minima. Bull. Jpn. Soc. Scient. Fish.* 42, 941, 1976.

Fusetani, N. and Hashimoto, Y. Hemolysins in a green alga *Ulva pertusa.* In, *Animal Plant and Microbial Toxins.* Ohsaka, A., Hayashi, K. and Sawai, Y. (eds.), Vol. 1, p. 325, Plenum Publishing Corp.: New York, 1976.

Gaillard, C. Recherches sur les poissons représentés dans quelques tombeaux Egyptiens de l' ancien empire. *Mém. Inst. Français Arch. Orient.* 51, 97, 1923.

Gallagher, J.P. and Shinnick-Gallagher, P. Effects of crude brevetoxin on membrane potential and spontaneous or evoked end-plate potentials in rat hemidiaphragm. *Toxicon* 23(3), 489, 1985.

Galsoff, P.S. Red tide progress report on the investigations of the cause of the mortality along the west coast of Florida conducted by the United States Fish and Wildlife Service and cooperating organizations. *Spec. Sci. Rep. U.S. Fish. Wildl. Serv.* 46, 1948.

Garber, P., Dutcher, J.D., Adams, E.G. *et al.* Protective effects of sea weed extracts for chicken embryos infected with influenza B or mumps virus. *Proc. Soc. Exp. Biol. Med.* 99, 590, 1958.

Garth, J.S. and Alcala, A.C. Poisonous crabs of Indo-West Pacific coral reefs, with special reference to the genus *Demania.* In, *Proc. Third Int. Coral Reef Symp.* Miami, 1977.

Gasteiger, E.L., Haake, P.C. and Gergen. J.A. An investigation of the distribution and function of homarine (N-methyl picolinic acid). *Ann. N.Y. Acad. Sci.* 90, 622, 1960.

Gaur, P.K., Anthony, R.L., Cody, T.S. *et al.* Production of a monoclonal antibody against sea nettle venom mouse lethal factor. *Proc. Soc. exp. Biol. Med.* 167, 374, 1981.

Gennaro, J.F., Lorincz, A.E. and Brewster, N.B. The anterior salivary gland of the octopus *(Octopus vulgaris)* and its mucous secretions. *Ann. N.Y. Acad. Sci.* 118, 1021, 1965.

Gershey, R.M., Neve, R.A., Musgrave, D.L. and Reichardt, P.B. A calorimetric method for the determination of saxitoxin. *J. Fish. Res. Bd. Can.* 34, 559, 1977.

Ghiretti, F. Cephalotoxin: the crab-paralysing agent of the posterior salivary glands of cephalopods. *Nature* 183, 1192, 1959.

Ghiretti, F. Toxicity of octopus saliva against crustacea. *Ann. N.Y. Acad. Sci.* 90, 726, 1960.

Ghosh, T.K., Gomes, A. and Nag Chaudhuri, A.K. Pharmacological action of tentacle extract of the jellyfish, *Acromitus rabanchatu,* occurring in the Bay of Bengal. *Ind. J. exp. Biol.* 28, 39, 1990a.

Ghosh, T.K., Ghosh, A. and Gomes, A. *et al.* Glycaemia effects of tentacle extract of the jellyfish *Acromitus rabanchatu* (Annandale). *Ind. J. exp. Biol.* 28, 441, 1990b.

Gibbs, F.J. and Greenaway, P. Histological structure of the posterior salivary glands in the blue-ringed octopus *Hapalochlaena maculosa. Toxicon* 16, 59, 1978.

Giebel, G.E.M., Thorington, G.U., Lim, R.Y. *et al.* Control of cnida discharge: II. Microbasic p-mastigophora nematocysts are regulated by two classes of chemoreceptors. *Biol. Bull.* 175, 132, 1988.

Giraldi, T., Ferlan, I. and Romeo, D. Antitumor activity of equinatoxin. *Chem.-biol. Interact.* 13, 199, 1976.

Giunio, P. Otrovne ribe. *Higijena* (Belgrade) 1, 282, 1948.

Glavin, G.B., Pinsky, C. and Bose, R. Mussel poisoning and excitatory amino acid receptors. *T.i.P.S.* 10, 5, 1990.

Gopinathan, C.P. On new distributional methods of plankton diatoms from the Indian seas. *J. Mar. Biol. Assoc. India* 17(1), 223–240, 1975.

Gorshkova, I.A., Gorshkova, B.A. and Stonik, V.A. Inhibition of rat brain Na^+-K^+-ATPase by triterpene glycosides from holothurians. *Toxicon* 27, 927, 1989.

Graham, A. Molluscan diets. *Proc. malac. Soc.* (Lond.) 31, 144, 1955.

Graneli, E., Sundstrom, B., Edler, L. *et al.* (eds.), *Toxic Marine Phytoplankton.* Elsvier: New York, 1990.

Grauer, F.H. Dermatitis escharotica caused by a marine alga. *Hawaii med. J.* 19, 32, 1959.

Grauer, F.H. and Arnold, H.L., Jr. Seaweed dermatitis. *Arch. Derm.* 84, 720, 1961.

Gravely, F.H. Shell and other animal remains found on the Madras beach. *Bull. Madras Govt. Mus.* 5, 86, 1941.

Gray, W.R., Olivera, B.M. and Cruz, L.J. Peptide toxins from venomous *Conus* snails. *Ann. Rev. Biochem.* 57, 665, 1988.

Green, G. Ecology of toxicity in marine sponges. *Mar. Biol.* 40, 207, 1977.

Grevin, J. *Deux Livres des Venins.* Christolfe Plantin: Paris, 1568.

Grevin, J. *De Venenis.* Plantin: Antwerp, 1571.

Grimmelt, B., Nijjar, M.S., Brown, J. *et al.* Relationship between domoic acid levels in the blue mussel *(Mytilus edulis)* and toxicity in mice. *Toxicon* 28(5), 501, 1990.

Grindley, J.R. and Taylor, F.J.R. Red water and mass mortality of fish near Cape Town. *Nature* (Lond.) 195, 1324, 1962.

Groweiss, A., Schmueli, U. and Kashman, Y. Marine toxins of *Latrunculia magnifica. J. org. Chem.* 48, 3512, 1983.

Gulavita, N.K., Scheuer, P.J. and deSilva, E.D. Antimicrobial constitutents of a spongenudibranch pair from Sri Lanka. In, *Bioactive Marine Compounds from Marine Organisms.* Thompson, M.F., Sarojni, R. and Nagabhushanam, R. (eds.), Oxford & IBH Publishing Co.: New Delhi, 1991.

Guldager, A. Doggerbankeeksem, forsøg pa profylaktisk behandling med corticosteroider. *Ugeskr. Laeg.* 121, 1567, 1959.

Gunthrope, L., and Cameron, A.M. Widespread but variable toxicity in scleractinian corals. *Toxicon* 28(10), 1199, 1990.

Gunthorpe, and Cameron, A.M. Intracolonial variation in toxicity in Scleractinian corals. *Toxicon* 28(10) 1221, 1990.

Gunthorpe, L., and Cameron, A.M. Toxic exudate from the hard coral *Goniopora tenuidens. Toxicon* 28(11), 1347, 1990

Habekost, R.C., Fraser, I.M. and Halstead, B.W. Observations on toxic marine algae. *J. Wash. Acad. Sci.* 45, 101, 1955.

Habermehl, G. and Volkwein, G. Über Gifte der mittelmeerischen Holothurien. II. Die Aglyka der Toxine von *Holothuria polii. Liebigs Ann. Chem.* 731, 53, 1970.

Habermehl, G. and Volkwein, G. Über Gifte der mittelmeerischen Holothurien. I. Mitteilung. *Naturwissenschaften* 55, 83, 1968.

Habermehl, G. and Volkwein, G. Aglycones of the toxins from the cuverian organs of *Holothuria forskäli* and a new nomenclature for the aglycones from Holothurioideae. *Toxicon* 9, 319, 1971.

Habermann, E. Palytoxin acts through Na$^+$, K$^+$ -ATPase. *Toxicon* 27, 1171, 1989.

Hach-Wunderle, V., Mebs, D., Frederking, K. *et al.* Vergiftung durch Feuerqualle. *Dtsch. Med. Wschr.* 112(48), 1865, 1987.

Haddock, R.L. and Cruz, O.L.T. Foodborne intoxication associated with seaweed. *The Lancet* 338, 195, 1991.

Halstead, B.W. *Poisonous and Venomous Marine Animals of the World.* 3 vols., U.S. Govt. Print. Off.: Washington, D.C., 1965–1970.

Halstead, B.W. *Poisonous and Venomous Marine Animals of the World.* Rev. edn. Darwin Press: Princeton, New Jersey, 1978.

Halstead, B.W. *Dangerous Marine Animals.* 2nd edn., p. 77, Cornell Maritime Press: Centreville, Maryland, 1980.

Halstead, B.W. *Poisonous and Venomous Marine Animals of the World.* Darwin Press: Princeton, New Jersey, 1988.

Halstead, B.W. and Haddock, R.L. A fatal outbreak of poisoning from the ingestion of red seaweed *Gracilaria tsudae* in Guam—review of the oral marine biotoxicity problem. *J. Nat. Toxins* 1(1), 87, 1992.

Halstead, B.W. and Russell, F.E. Animal Toxins. In, *Biology Data Book.* Altman, P.L. and Dittmar, D.S. (eds.). FASEB: Washington, D.C., 1964.

Hand, C. Present state of nematocyst research: types, structure and function. In, *The Biology of Hydra and of Some Other Coelenterates.* Lenhoff, H.M. and Loomis, W. F., (eds.), University of Miami Press: Coral Gables, Floriba, 1961.

Hartman, W.J., Clark, W.G., Cry, S.C. *et al.* Pharmacologically active amines and their biogenesis in the octopus. *Proc. West. pharmacol. Soc.* 3, 106, 1960.

Hashimoto, Y. *Marine Toxins and Other Bioactive Metabolites.* Japan Scientific Society: Tokyo, 1979.

Hashimoto, Y. and Ashida, K. Screening of toxic corals and isolation of a toxic polypeptide from *Goniopora* spp. *Pulb. Seto Mar. Biol. Lab.* 20, 703, 1973.

Hashimoto, Y., Fusetani, N. and Kimura, S. Aluterin: a toxin of file fish, *Alutera scripta*, probably originating from a zoantharian, *Palythoa tuberculosa.* *Bull. Jpn. Soc. Scient. Fish.* 35, 1086, 1969.

Hashimoto, Y., Konosu, S., Yasumoto, T. *et al.* Occurrence of toxic crabs in Ryuku and Amami Islands. *Toxicon* 5, 85, 1967.

Hashimoto, Y., Miyazawa, K., Kamiya, H. *et al.* Toxicity of the Japanese ivory shell. *Bull. Jpn. Soc. scient. Fish.* 33, 661 1967a.

Hasimoto, Y., Naito, K. and Tsutumi, J. Photosensitization of animals by the viscera of abalones, *Haliotis* spp. *Bull. Jpn. Soc. scient. Fish.* 26, 1216, 1960.

Hasimoto, Y. and Yasumoto, T. Confirmation of saponin as a toxic principle of starfish. *Bull. Jpn. Soc. scient. Fish.* 26, 1132, 1960.

Hauano, Y., Kinoshita, Y. and Yasumoto, T. Suckling mice assay for Diarretic shellfish toxins. In, *Toxic Dinoflagellates.* Anderson, M., White, A.W. and Baden, D.G. (eds.) p. 381, Elsevier: New York, 1985.

Hellberg, S. and Kem, W.R. Quantitative structure-activity relationships for sea anemone polypeptide toxins. *Int. J. Peptide Protein Res.* 36, 440, 1990.

Hemingway, G.T. Evidence for a paralytic venom in the intertidal snail, *Acanthina spirata* (Neogastropoda-Thaisidae). *Comp. Biochem. Physiol.* 60C, 79, 1978.

Hemmert, W.H. The public health implications of *Gymndinium breve* red tides: A review of the literature and recent events. In, *Ist International Conference on Toxic Dinoflagellate Blooms.* Lociero, U.R. (ed.), *Mass. Sci. Tech. Found.*: Wakefield, Mass., p. 489, 1975.

Henri, V. and Kayalof, E. Etude des toxines contenues dans les pédicellaires chez les Oursins. *C.R. Soc. Biol.* 60, 884, 1906.

Hermitte, L.C. Venomous marine molluscs of the genus *Conus. Trans. Roy. Soc. trop. Med. Hyg.* 39, 485, 1946.

Hessinger, D.A. and Ford, M.T. Ultrastructure of the small cnidocyte of the Portuguese man-of-war (*Physalia physalis*) tentacle. In, *The Biology of Nematocysts.* Hessinger, D.A. and Lenhoff, H.M. (eds.), p. 75, Academic Press: Orlando, 1988.

Hessinger, D.A. and Lenhoff, H.M. Membrane structure and function. Mechanism of hemolysis induced by nematocyst venom: roles of phospholipase and direct lytic factor. *Arch. Biochem. Biophys.* 173, 603, 1976.

Hillard, D.W. Science and scientists. *Med. Arts and Sciences* 3,1,1938.

Hinegardner, R.T. The venom apparatus of the cone shell. *Hawaii med. J.* 17, 533, 1958.

Hinegardner, R.T. *Morphology of the Venom Apparatus of Conus.* Thesis, Univ. S. California: Los Angeles, 1957.

Hines, K. and Lane, C.E. Amino acid compostion of active peptides of *Physalia* toxin. *Fed. Proc.* 21, 53, 1962.

Hirata, Y., Uemura, D. and Ohizumi, Y. Chemistry and pharmocology of palytoxin. In, *Handbook of Natural Toxins.* Tu, A.T. (ed.), vol. III, p. 241. Marcel Dekker: New York, 1988.

Hirata, Y., Uemura, D., Ueda, K. et al. Several compounds from *Palythoa tuberculosa* (Coelenterata). *Pure Appl. Chem.* 51, 1875, 1979.

Hirayama, H., Gohgi, K., Urakawa, N. *et al.* A ganglionic-blocking action of the toxin isolated from the Japanese ivory shell (*Babylonia japonica*). *Jpn. J. Pharmac.* 20, 311, 1970.

Ho, C.L. and Hwang, L.L. Isolation and biological assay of gonyautoxins from the cultured purple clam (*Coletellina diphos*). *J. Chin. Biochem. Soc.* 18(2), 56, 1989.

Ho, C.L., Ko, J.L., Lue, C.M. *et al.* Effects of equinatoxin on the guinea-pig atrium. *Toxicon* 25, 659, 1987.

Hokama, Y. Immunological analysis of low molecular weight marine toxins. *J. Toxicol.-Toxin Rev.* 10, 1, 1991.

Holstein, T. and Tardent, P. An ultrahigh-speed analysis of exocytosis: nematocyst discharge. *Science* 223, 830, 1984.

Hong, S.J. and Chang, C.C. Use of geographic-toxin II (μ-conotoxin) for the study of neuromuscular transmission in mouse. *Br. J. Pharmacol.* 97, 934, 1989.

Hooper, J.N.A., Capon, R.J. and Hodder, R.A. A new species of toxic marine sponge (Porifera: Desmospongiae: Poecilosclerida) from northwest Australia. *The Beagle* (Records of Northern Terr. Mus. Arts Sci.) 8(1), 27, 1991.

Hopkins, D.G. Venomous effects and treatment of octopus bite. *Med. J. Aust.* (3), 81, 1964.

Hoppe, H.A., Levring, T. and Tanaka, Y. *Marine Algae in Pharmaceutical Science.* Walter de Gruyter: Berlin/New York, 1979.

Hoppe-Seyler, F.A. Uber das Homarin, eine bisher unbekannte tierishe Base. *Hoppe-Seyler Z.* 222, 105, 1933.

Hori, A. and Shimizu, Y. Biosynthetic N-enrichment and N-nmr spectra of neosaxitoxin and gonyautoxin-II: application to the structure determination. *J. Chem. Soc. Chem. Commun.* ?, 790, 1983.

Hornell, J. A new protozoan cause of widespread mortality among marine fishes. *Bull. Madras Fish. Dept.* 11, 53, 1917.

Howden, M.E.H. and Williams, P.A. Occurrence of amines in the posterior salivary glands of the octopus *Hapalochlaena maculosa* (Cephalopoda). *Toxicon* 12, 317, 1974.

Hulme, F.E. *Natural History of Lore and Legend.* 350 pp. London, 1895.

Huang, C.L. and Mir, G.N. Pharmacological investigation of salivary gland of *Thais haemastoma* (Clench). *Toxicon* 10, 111, 1972.

Huot, R.I., Armstrong, D.L. and Chanh, T.C. Protection against nerve toxicity by monoclonal antibodies to sodium channel blocker tetrodotoxin. *J. Clin. Invest.* 83, 1821, 1989.

Hurst, J.W., Selvin, R. and Sullivan, J.L. Intercomparison of various assay methods for the detection of shellfish toxins. In, *Dinoflagellate Toxins.* Andrerson, M., White, A.W. and Baden, D.G. (eds.), p. 427. Elsevier: New York, 1985.

Hyman, L.H. *The Invertebrates: Protozoa Through Ctenophora.* McGraw-Hill: New York, 1940.

Ikegami, S., Kamiya, Y. and Tamura, S. Isolation and characterization of spawning inhibitors in ovaries of the starfish, *Asterias amurensis. Agric. Biol. Chem.* 36, 1087, 1972.

Ilangovan, G. Studies on dinoflagellates from Porto Novo regions, India. Ph.D. thesis, Annamalai University, 1981.

Imai, M. and Inoue, K. The mechanism of the action of prymnesium toxin on membranes. *Biochim. Biophys. Acta* 352, 344, 1974.

Inoue, A., Naguchi, T., Konosu, S. *et al.* A new toxic crab, *Atergatis floridus. Toxicon* 6, 119, 1968.

Ioannides, G. and Davis, J.H. Portuguese man-oı-war stinging. *Arch. Derm* 91, 312, 1965.

Isgrove, A. *Eledone. Mem. Liverpool Mar. Biol. Comm.* 16, 105, 1909.

Ishidate, M. and Hagiwara, H. On the poisonous substance isolated from asari (*Venerupis semidcussata*) and oyster (In Japanese). *Exp. Rept. Dept. Nat. Sci. Univ. Tokyo* 10, 43, 1954.

Irie, T., Suzuki, M., Hayakawa, Y. Isolation of aplysin, debromoaplysin, and aplysinol. *Bull. Chem. Soc. Jpn.* 42, 843, 1969.

Iverson, F., Truelove, J., Nera, E. *et al.* Domoic acid poisoning and mussel associated intoxication: preliminary investigations into the response of mice and rats to toxic mussel extract. *Fd. Chem. Toxic.* 27, 377, 1989.

Jakowska, S. and Nigrelli, R.F. Antimicrobial substances from sponges. *Ann. N.Y. Acad. Sci.* 90, 913, 1960.

James, D.B., Vivekanandam, E. and Srinivasarengan, S. Menace from medusae off Madras with notes on their utility and toxicity. *J. Marine biol. Assoc. India* 27(1/2), 170, 1985.

Jarvis, M.W., Crone, H.D., Freeman, S.E. *et al.* Chromatographic properties of maculotoxin, a toxin secreted by *Octopus (Hapalochlaena) maculosus. Toxicon* 13, 177, 1975.

Jayasree, V., Gupta, R.S. and Bhavanarayana, P.V. A toxin from *Holothuria leucospilota* (Brandt). *Nat. Inst. Ocean,* Goa, 1989.

Jennings, R.K. and Aker, R.F. The *Protistan Kingdom.* Van Nostrand Reinhold: New York, 1970.

Johannsen, J.E., Svec, W.A., Liaaen-Jensen, S. *et al.* Carotenoids of the Dinophyceae. *Phytochem.* 13, 2261, 1974.

Johnson, H.M., Frey P.A. *et al.* Haptenic properties of paralytic shellfish poisoning. *Proc. Soc. exp. Biol. Med.* 117, 425, 1964.

Johnson, G.R., Hanic, L., Judson, I. *et al.* Mussel culture and the accumulation of domoic acid. *Can. Dis. Wkly. Rept.* Suppl. 1E, 33, 1990.

Jullien, A. Variations dans le temps de la teneur des extraits de glande à pourpre en substances actives sur le muscle de sangsue. *C.R. Soc. Biol.* 133, 524, 1940.

Kaiser, E. and Michi, H. *Die Biochemie der tierischen Gifte.* Franz Deuticke: Wien, 1958.

Kao, C.Y. Comparison of the biological actions of tetrodotoxin and saxitoxin. In, *Animal Toxins.* Russell, F.E. and Saunders, P.R. (eds.), p. 109 Pergamon: Oxford. 1967.

Kao, C.Y. and Walker, S.E. Active groups of saxitoxin and tetrodotoxin as deduced from action of saxitoxin analogues on frog muscle and squid axon. *J. Physiol.* 323, 619, 1982.

Karunasagar, I. Outbreak of paralytic shellfish poisoning in Mangalore, west coast of India. *Curr. Sci.* 53(5), 247, 1984.

Karunasagar, I., Gowda, H.S.V., Subbaraj, M. *et al.* Outbreak of paralytic shellfish poisoning in Mangalore, west coast of India. *Curr. Sci.* 53(5), 247, 1984.

Karunasagar, I., Segar, K. and Karunasagar, I. Potentially toxic dinoflagellates in shellfish harvesting areas along the coast of Karnataka State (India). In, *Red Tides: Biology.* Okaichi, T., Anderson, D.M. and Nemoto, T. (eds.), Elsevier: New York, 1989.

Karunasagar, I. and Segar, K. Incidence of PSP and DSP in shellfish along the coast of Karnataka State (India). In, *Red Tides: Biology Environmental Sciences and Technology.* Okaichi, T., Anderson, D.M. and Nemoto, T., (eds.), p. 61, Elsevier: New York, 1989.

Karunasagar, I., Karunasagar, I., Oshima, Y. and Yasumoto, T. A toxin profile in shellfish involved in an outbreak of paralytic shellfish poisoning in India. *Toxicon* 28, 868, 1990.

Kashman, Y., Fishelson, L. and Neéman, I. N-acyl-2-methylene-β-alanine methyl esters from the sponge *Fasciospongia cavernosa. Tetrahedron* 29, 3655, 1973.

Kasinathan, R., Chandrasekaran, U.S. and Tagore, J. Effects of *Conus amidas* toxin on the hematological variables in the estuarine fish, *Mugil cephalus. Toxicon* 27, 53, 1989; see also 54, 1989.

Kasinathan, R., Chandrasekaran, V.S. and Tagore, J. Effects of *Conus aulicus* toxin on the body component indices of three estuarine fishes. *Toxicon* 27, 54, 1989.

Kasinathan, R., Ramu, Y.D. Ravichandran, V. *et al.* Pharmacological studies on *Conus bayani* from southeast coast of India. *10th World Congress on Animal Plant and Micronial Toxins.* Nov. 1991, Singapore. Abs. 218, p. 270.

Kato, Y. and Scheuer, P.J. Aplysiatoxin and debromoaplysiatoxin, constituents of the marine mollusc *Stylocheilus longicauda* (Quoy and Gaimard, 1824). *J. Am. chem. Soc.* 96, 2245, 1974.

Kato, Y. and Scheuer, P.J. The aplysiatoxins. *Pure appl. Chem.* 42, 1, 1975.

Kaul, P. Biomedical potential of the sea. *Pure appl. Chem.* 54, 1963, 1982.

Kawabata, T., Halstead, B.W. and Judefind, T.F. A report of a series of recent outbreaks of unusual cephalopod and fish intoxication in Japan. *Am. J. trop. Med. Hyg.* 6, 935, 1957.

Keen, T.E.B. and Crone, H.D. The hemolytic properties of extracts of tentacles from the cnidarian *Chironex fleckeri. Toxicon* 7, 55, 1969a.

Keen, T.E.B. and Crone, H.D. Dermonecrotic properties of extracts from the tentacles of the cnidarian *Chironex fleckeri. Toxicon* 7, 143, 1969b.

Kelecom, A., Daloze, D. and Tursch, B. Chemical studies of marine invertebrates. *Tetrahedron* 32, 2313, 1976.

Kellaway, C.H. The action of mussel poison on the nervous system. *Aust. J. exp. Biol. Med. Sci.* 13, 79, 1935a.

Kellaway, C.H. Mussel poisoning. *Med. J. Aust.* (1), 399, 1935b.

Kelman, S.N., Calton, G.J. and Burnett, J.W. Isolation and partial characterization of a lethal sea nettle (*Chrysaora quinquecirrha*) mesenteric toxin. *Toxicon* 22(1), 139, 1984.

Kem, W.R.. *A Chemical Investigation of Nemertine Toxins.* Ph.D. Thesis, University of Illinois: Urbana, Illinois, 1969.

Kem, W.R. A study of the occurrence of anabaseine in *Paranemertes* and other nemertines. *Toxicon* 9, 23, 1971.

Kem, W.R. Biochemistry of nemertine toxins. In, *Marine Pharmacognosy. Action of Marine Biotoxins at the Cellular Level.* Martin, D.F. and Padilla, G.M. (eds.), Academic Press: New York and London, 1973.

Kem, W.R. Purification and characterization of a new family of polypeptide neurotoxins from the heteronemertine *Cerebratulus lacteus* (Leidy). *J. Biol Chem.* 251(14), 4184, 1976.

Kem, W.R. Sea anemone toxins: structure and function. In, *Biology of Nematocysts.* Hessinger, D. and Lenhoff, H. (eds.), Academic Press: London, 1988.

Kem, W.R., Abbott, B.C. and Coates, R.M. Isolation and structure of a hoplonemertine toxin. *Toxicon* 9, 15, 1971.

Kem, W.R. and Blumenthal, K.M. Purification and characterization of the cytolytic *Cerebratulus* A toxins. *J. biol. Chem.* 253, 5752, 1978a.

Kem, W.R. and Blumenthal, K.M. Polypeptide cytolysins and neurotoxins isolated from the mucus secretions of the heteronemertine *Cerebratulus lacteus* (Leidy). In, *Toxins, Animal, Plant and Microbial.* Resenberg, P. (ed.), Pergamon Press: Oxford, 1978b.

Kem, W.R., Blumenthal, K.M. and Doyle, J.W. Cytotoxins of some marine invertebrates. In, *Natural Toxins.* Eaker, D. and Wadström, T. (eds.), Pergamon Press: Oxford, 1980.

Kem, W.R., Dunn, B.M., Paraten, B. *et al.* The unconventional amino acid sequence of the major neurotoxic polypeptide of the sun anemone *Stoichactis helianthus. Toxicon* 23, 580, 1985.

Kem, W.R., Scott, K.N. and Duncan, J.H. Hoplonemertine worms—a new source of pyridine neurotoxins. *Experientia* 32, 684, 1976.

Kendrew, J.C., Dickerson, R.E., Strandberg, B.E. *et al.* Structure of myoglobin. *Nature* 185, 422, 1980.

Keyl, M.J., Michaelson, I.A. and Whittaker, V.P. Physiologically active choline esters in certain gastropods and other invertebrates. *J. Physiol.* 139, 434, 1957.

Kim, Y.S. and Padilla, G.A. Hemolytically active components from *Prymnesium parvum* and *Gymnodinium breve* toxins. *Life Sci.* 21, 1287, 1977.

Kimura, A., Hayashi, H. and Kuramoto M. Studies of urchi-toxins: separation, purification and pharmacological actions of toxinic substances. *Jpn. J. Pharmacol.* 25, 109, 1975.

Kimura, A., Nakayawa, H., Hayashi, H. *et al.* Seasonal changes in contractile activity of a toxic substance from the pedicellariae of the sea urchin *Toxopneustes pileolus*. *Toxicon* 22(3), 353, 1984.

King. H. Amphiporine, an active base from the marine worm *Amphiporus lactifloreus*. *J. Am. chem. Soc.* 2, 1365, 1939.

Kinnel, R., Duggan, A.J., Eisner, T. et al. Panacene: an aromatic broncalene from a sea hare *(Aplysia brasiliana)*. *Tetrahed. Lett.* 44, 3913, 1977.

Kitagawa, I., Sugaware, T. and Yoshioka, I. *et al.* Structure of Holotoxin A, a major antifungal glycoside of *Stichopus japonicus* Selenka. *Tetrahedron Lett.* 44, 4111, 1974.

Kitagawa, I., Sugaware, T. and Yoshioka, I. [Holotoxins of *Stichopus* sp.] *Chem. Pharm. Bull.* 24, 275, 1976.

Kittredge, J.S., Takahashi, F.T., Lindsey, J. *et al.* Chemical signals in the sea: Marine allelochemics and evolution. *Fish. Bull.* 72, 1, 1974.

Klauber, L.M. *Rattlesnakes.* University of California Press: Berkeley, California, 1956.

Kleinhaus, A.L., Cranefield, P.F. and Burnett, J.W. The effects on canine cardiac Purkinje fibers of *Chrysaora quinquecirrha* (sea nettle) toxin. *Toxicon* 11, 341, 1973.

Kodama, A.M., Hokama, Y. and Yasumoto, T. Clinical and laboratory findings implicating palytoxin as cause of ciguatera poisoning due to *Decapterus macrosoma* (mackerel). *Toxicon* 27, 1051, 1989.

Kodama, M., Ogata, T. and Sato, S. Bacterial production of saxitoxin. *Agri. Biol. Chem.* 52, 1975, 1988.

Kodama, M., Ogata, T., Sakamoto, S. *et al.* Production of paralytic shellfish toxins by a bacterium *Moraxella* sp. isolated from *Protogonyaulax tamarensis*. *Toxicon* 28, 707, 1990.

Kofold, C.A. Dinoflagellata of the San Diego region. IV. The genus *Gonyaulax*, with notes on its skeletal morpholgy and a discussion of its generic and specific characterisitics. *Univ. Calif. Zool.* 8, 187, 1911.

Kohn, A.J., Saunders, P.R. and Wiener, S. Preliminary studies on the venom of the marine snail *Conus. Ann. N.Y. Acad. Sci.* 90, 76, 1960.

Kokelj, F. and Burnett, J.W. Treatment of a pigmental lesion induced by a *Pelagia noctiluca* sting. *Cutis* 46, 62, 1990.

Konishi, K. Studies on organic insecticides. Part XII. Synthesis of nereistoxin and related compounds. *V. Agric. biol. Chem.* 34, 935, 1970.

Konosu, S., Inoue, A., Noguchi, T. *et al.* Comparison of crab toxin with saxitoxin and tetrodotoxin. *Toxicon* 6, 113, 1968a.

Konosu, S., Inoue, A., Noguchi, T. *et al.* On the venomous crustacea and fish from Amami-Oshima and Okinawa Islands. I. *Med. J. Kagoshima Univ.* 19, 729, 1968b.

Kosuge, T., Zenda, H., Ochiai, A. *et al.* Isolation and structure determination of a new marine toxin, surugatoxin, from the Japanese ivory shell *Babylonia japonica. Tetrahed. Lett.* 26, 2545, 1972.

Kotaki, Y., Oshima, Y. and Yasumoto, T. Bacterial transformation of paralytic shellfish poisons. In, *Toxic Dinoflagellates.* Anderson, D.M., White A.W. and Baden, D.G. (eds.), p. 387, Elsevier: New York, 1985.

Kreger, A.S. Cytolytic activity in the venom of the marine snail *Conus californicus. Toxicon* 27, 56, 1989.

Kuznetsova, T.A., Anismov, M., Popov, A.M. *et al.* A comparative study *in vitro* of physiological activity of triterpene glycosides of marine invertebrates of echinoderm type. *Comp. Biochem. Physiol.* 73 C (1), 41, 1982.

Lafranconi, W.M., Ferlan, I., Russell, F.E. *et al.* The action of equinatoxin, a peptide from the venom of the sea anemone, *Actinia equina. Toxicon* 22, 347, 1984.

Lal, D.M. Calton, G.J., Neéman. I. *et al.* Characterization of *Chrysaora quinquecirrha* (sea nettle) nematocyst venom collagenase. *Comp. Biochem. Physiol.* 69B, 529, 1981.

Lane, C.E. The toxin of *Physalia* nematocysts. *Ann. N.Y. Acad. Sci.* 90, 742, 1960.

Lane, C.E. *Physalia* nematocysts and their toxins. In, *Biology of Hydra.* Lenoff and Lomis (eds.), Miami Press: Coral Gables, Florida, 1961.

Lane, C.E. Coelenterata: chemical aspects of ecology, pharmacology and toxicology. In, *Chemical Zoology.* Florkin, M. and Scheer, B.T. (eds.), Academic Press: New York, 1968.

Lane, C.E. and Dodge, E. The Toxicity of *Physalia* nematocysts. *Biol. Bull.* 115, 219, 1958.

Lane, C.E. and Larsen, J.B. Some effects of the toxin of *Physalia physalis* on the heart of the land crab, *Cardisoma guanhumi* (Latreille). *Toxicon* 3, 69, 1965.

Larsen, J.B. and Lane, C.E. Some effects of *Physalia physalis* toxin on the cardiovascular system of the rat. *Toxicon* 4, 199, 1966.

Larsen, J.B. and Lane, C.E. Direct action of *Physalia* Toxin on frog nerve and muscle. *Toxicon* 8, 21 1970.

Lau, C.O., Chng, S.Y., Chia, D.G.B. *et al.* A preliminary survey of local marine crabs for toxicity. *10th World Congress on Animal, Plant and Microbial Toxins.* Singapore, Nov. 1991, Abs. 221, p. 274.

Lee, J.S., Yanagi, T. and Kenma, R. *et al.* Fluorometric determination of diarrhetic shellfish toxins by high-performance liquid chromatography. *Agric. Biol. Chem.* 51, 877, 1987.

LeMare, D.W. Poisonous Malayan fish. *Med. J. Malaya* 7, 1, 1952.

Lenhoff, H.M., Kline, E.S. and Hurley, R. A hydroxyproline-rich, intracellular, collagen-like protein of *Hydra* nematocysts. *Biochim. Biophys. Acta* 26, 204, 1957.

Lentz, T.L. Histochemical localization of neurohumors in a sponge. *J. exp. Zool.* 162, 171, 1966.

Lentz, T.L. and Barnett, R.J. Fine structure of the nervous system of *Hydra. Amer. Zool.* 5, 341, 1965.

Levine. L., Fujiki, H., Gjika, H.B. *et al.* A radioimmunoassay for palytoxin. *Toxicon* 26, 1115, 1988.

Levy, R. Sur les properiétés hémolytiques des pédicellaires de certains oursins réguliers. *C.R. Acad. Sci.* 181, 690, 1925.

Lewis, J., Tett, P., and Dodge, J.D. The syst-theca cycle of *Gonyaulax polyedra (Lingulodinium machaerophorum)* in Creran, a Scottish west coast sea-loch. In, *Toxic Dinoflagellates.* Anderson, D.M., White, A.W. and Baden, D.G. (eds.), p. 85, Elsevier: New York, 1985.

Liebert, F. and Deerns, W., Ondeszock naar de oorzaak van den vischsterfte in den polur Workmumer-Nieuwland, nabij Workum. *Verh. Rapp. Rijksinst. Visscherijonderzoe* 1, 81, 1920.

Lightner, D.V. Possible toxic effects of the marine blue-green alga. *Spherulina subsala*, on the blue shrimp. *Penareus stylirostris. J. Invert. Path.* 32, 139, 1978.

Lim, H.H., Tan, C.K., Cha, S.W. *et al.* Effects of intracisternal acid intraneural injection of the purified toxin of a local coral reef crab, *Lophozozymus pictor. 10th World Congress on Animal, Plant and Microbial Toxins.* Nov. 1991, Singapore. Abs. 222, p. 275.

Lindner, G. Über giftige Miesmuscheln. *Centralbl. Zent. Bak. Parasiten* 3, 352, 1888.

Llewellyn, L.E. and Endean, R. Toxins extracted from Australian specimens of the crab *Eriphia sebana* (Xanthidae). *Toxicon* 27, 579, 1989.

Llewellyn, L.E. and Endean, R. Toxicity and paralytic shellfish toxin profiles of the xanthid crabs *Lophozozymus pictor* and *Zosimus aeneus*, collected from South Australian coral reefs. *Toxicon* 27, 596, 1989.

LoBianco, S. Notizie biologiche rigurdanti specialmente il periodo di maturita sessuale degli animali del Golfo di Napoli. *Mitt. Zool. Sta. Neapel.* 8, 385, 1888.

Loisel, G. Les poisons des glandes génitales. 1er note: recherches et expérimentation chez l'oursin. *C.R. Soc. Biol.* 55, 1329, 1903.

Loisel, G. Recherches sur les poisons génitaux de différents animaux. *C.R. Acad. Sci.* 139, 227, 1904.

Lubbock, R. Chemical recognition and nematocyte excitation in a sea anemone. *J. exp. Biol.* 83, 283, 1979.

Lubbock, R. Why are clownfishes not stung by sea anemones? *Proc. Roy Soc. London* B207, 35, 1980.

Mabbet, H. Death of a skindiver. *Austral. Skin Div. Spear Fish. Dig.* (Dec.), p. 13, 17, 1954.

Macfarlane, R.D., Uemura, D. and Hirata, Y. Cf plasma desorption mass spectrometry of palytoxin. *J. Am. Chem. Soc.* 102, 875, 1980.

MacGinitie, G.E. Notes on the natural history of some marine animals. *Am. Midland Naturalist* 19, 213, 1942.

Macht, D.I., Brooks, D.J. and Spencer, E.C. Physiological and toxicological effects of some fish muscle extracts. *Am J. Physiol.* 133, 372, 1941.

Mačkie, A.M. and Turner, A.B. Partial characterization of a biologically active steroid glycoside isolated from starfish *Marthasterias glacialis. Biochem. J.* 117, 543, 1970.

Mačkie, G.O. Studies on *Physalia physalis* (L.) Part 2. Behavior and histology. *Discovery Rep.* 30, 371, 1960.

Maeda, M., Kodama, T., Tanaka, T. *et al.* Structure of insecticidal substances isolated from a red alga, *Chondria aomata. Tennen Yuki Kagobutsu Toronkai Koen Yoshishu* 27, 616, 1985.

Maiti, B.C., Thomson, R.H. and Mahendra, M. Structure of caulerpin, a pigment from *Caulerpa algae. J. chem. Res.* 4, 126, 1978.

Maček, P. and Lebez, D. Kinetics of hemolysis induced by equinatoxin, cytolytic toxin from the sea anemone *Actinia equina.* Effects on some ions and pH. *Toxicon* 19, 233, 1981.

Maček, P. and Lebez, D. Isolation and charcterization of three lethal and hemolytic toxins from one sea anemone *Actinia equina* L. *Toxicon* 26, 441, 1988.

Mackie, A.M. and Turner, A.B. Partial characterization of a biologically active steroid glycoside isolated from starfish *Marthasterias glacialis. Biochem. J.* 117, 543, 1970.

Mackie, A.M., Lasker, R. and Grant, P.T. Avoidance reactions of the molluse *Buccinum undatum* to saponin-like surface-active substances in extracts of the starfish *Asterias rubens* and *Marthasterias glacialis. Comp. Biochem. Physiol.* 26, 415, 1968.

Maclean, J.L. and White, A.W. Toxic dinoflagellate blooms in Asia: a growing concern. In, *Toxic Dinoflagellates.* Anderson, D.M., White, A.W. and Baden, D.G. (eds.), p. 517, Elsevier: New York, 1985.

Maranda, L., Anderson, D.M. and Shimizu, Y. Comparison of toxicity between populations of *Gonyaulax tamarensis* of eastern North American waters. *Estuarine. Coastal Shelf Sci.* 21, 401, 1985.

Maretić, Z. and Russell, F.E. Stings by the anemone *Anemonia sulcata. Am. J. trop. Med. Hyg.* 32, 891, 1983.

Mariscal, R.N. Nematocysts. In, *Coelenterate Biology.* Muscatine, L. and Lenhoff, H.M. (eds.), Academic Press: London and New York, 1974.

Marsh, H. The caseinase activity of some vermivorous Conidae. *Toxicon* 8, 271, 1971.

Marshall, I.G. and Harvey, A.L. Selective neuromuscular blocking properties of α-conotoxins *in vivo. Toxicon* 28(2), 231, 1990.

Maratin, J.C. and Audley, I. Cardiac failure following Irukandji envenomation. *Med. J. Aust.* 153, 164, 1990.

Martin, E.J. Observations on the toxic sea anemone *Rhodactis howesii (Coelenterata). Pac. Sci.* 14, 403, 1960.

Martin, D.F. and Chatterjee, A.B. Some chemical and physical properties of two toxins from red-tide organism, *Gymnodinium breve. U.S. Fish Wildl. Serv., Fish. Bull.* 68, 433, 1970.

Mas, R., Menéndez, R. Garateix, A. *et al.* Effects of a high molecular weight toxin from *Physalia physalis* on glutamate responses. *Neuroscience* 33(2), 269, 1989.

Matsuda, H. and Tomiue, Y. The structure of aplysin-20. *Chem. Commun.* 898, 1967.

Matsumo, T. and Iba T. [Studies on the saponin of sea cucumber.] *J. pharm. Soc. Jpn.* 86, 637, 1966.

Matsumo, T. and Yamanouchi, T. A. new triterpenoid sapogenin of animal origin (sea cucumber). *Nature* 191, 75, 1961.

Matus, A.I. Fine structure of the posterior salivary gland of *Eledone cirrosa and Octopus vulgaris. Z. Zellforsch.* 122., 11, 1971.

McClean, J.L. In, *Toxic Red Tides and Shellfish Toxicity in Southeast Asia.* White, A.W., Anraku, M. and Hooii, K.K. (eds.), S.E. Asian Fish. Development Center and Intern. Deve. Res. Center: Singapore, 1984.

McFarren, E.F., Schafer, M.L., Campbell, *et al.* Public health significance of paralytic shellfish poison. A review of literature and unpublished research. *Proc. Natn. Shellfish Ass.* 47, 114, 1956.

McFarren, E.F., Schafer, M.L., Campbell, J.E. *et al.* Public health significance of paralytic shellfish poison. *Adv. Food. Res.* 10, 135, 1960.

McMichael, D.F. Dangerous marine molluscs. *Proc.* Ist. Intern. Convention Life Saving Techniques, Part III, Sci. Sect., p. 86. Bull. Post Grad. Comm. Med., Univ. Sydney, Australia, 1963.

Mebs D. and Gebauer, E. Isolation of proteinase inhibitor, toxic and hemolytic polypeptides from a sea anemone *Stoichactis helianthus. Toxicon* 18, 97, 1980.

Mebs, D., Weiber, I. and Heineke, H.F. Bioactive proteins from marine sponges: screening of sponge extracts fro hemagglutinating, hemolytic, icthyotoxic and lethal properties and isolaton and characterization of hemagglutinins. *Toxicon* 23, 955, 1985.

Medcof, J.C., Leim, A.H., Needler, A.B. *et al.* Paralytic shellfish poisoning on the Canadian Atlantic coast. *Bull. Fish. Res. Bd. Can.* 75, 1, 1947.

Mendes, E.G. Ulteriores experimentos sobre o principio colinergico em pedicellarias. *Cienc. e Cult.* 15, 275, 1963.

Menon, M.G.K. The scyphomedusae of Madras and neighbouring coast. *Bull. Madras Govt. Mus. N.S.* (Nat. Hist.) 3, 1, 1930.

Meyer, K.F. Medical progress: food poisoning. *New Engl. J. Med.* 249, 765, 804, 843, 1953.

Meyer, K.F., Sommer, H. and Schoenholz, P. Mussel poisoning. *J. Prev. Med.* 2, 356, 1928.

Michaels, W.D. Membrane damage by a toxin from the sea anemone *Stoichactis helianthus-I.* Formation of trans membrane channels in lipid bilayers. *Biochim. Biophys. Acta* 555, 67, 1979.

Michel, C. and Keil. B. Biologically active proteins in the venomous glands of the polychaetous annelid, *Glycera convoluta* Keferstein. *Comp. Biochem. Physiol.* 50B, 29, 1975.

Middlebrook, R.E., Wittle, L.W. Scura, E.D. et al. Isolation and purification of a toxin from *Millepora dichotoma. Toxicon* 9, 333, 1971.

Miller, D.M. and Tindall, D.R. Physiological effects of HPLC-purified maitotoxin from dinoflagellate, *Gambierdiscus toxicus.* In, *Toxic Dinoflagellates.* Anderson, D.M., White, A.W. and Baden, D.G. (eds.), p. 375, Elsevier: New York, 1985.

Mold, J.D., Bowden, J.P., Stanger, D.W. *et al.* Paralytic shellfish poison: VII. Evidence for the purity of the poison isolated from toxic clams and mussels. *J. Am. chem. Soc.* 79, 5235, 1957.

Möller, H. and Beress, L. Effect on fishes of two toxic polypeptides isolated from *Anemonia sulcata. Mar. Biol.* 32, 189, 1975.

Montgomery, D.H. Responses of two haliotid gastropods (Mollusca) *Haliotis asimilis and Haliotis rufescens* to the forcipulate asteroids (Echinodermata) *Pycnopodia helianthoides* and *Pisaster ochraceus. Veliger* 9, 359, 1967.

Moore, R.E. Toxin from the blue-green algae. *Bioscience* 27, 797, 1977.

Moore, R.E. and Bartolini, G. Structure of palytoxin. *J. Am. Chem. Soc.* 103, 2491, 1981.

Moore, R.E. Dietrich, R.F., Hatton, B. *et al.* The nature of the 263 chromophore in the palytoxins. *Org. Chem.* 40, 540, 1975.

Moore, R.E. and Scheuer, P.J. Palytoxin: a new marine toxin from a coelenterate. *Science*. 172, 495, 1971.

Mori, Y., Anraku, M., Yagi, K. *et al*. On the venomous crustacea and fish from Amami-Oshima and Okinaawa Islands. I. *Med. J. Kagoshima Univ*. 19, 729, 1968.

Mortensen, T. *A Monograph of Echinoidea*. 5 vols., C.A. Retizel: Copenhagen, 1928-1951.

Muhvich, K.H., Sengottuvelu, S., Manson, P.N, *et al*. Pathophysiology of sea nettle (*Chrysaora quinquecirrha*) envenomation in a rat model. *Toxicon* 29(7), 857, 1991.

Müller, H. Chemistry and toxicity of mussel poison. *J. Pharmacol. Exp. Ther*. 53, 67, 1935.

Munro, H.S. *The Nature and Mode of Action of Coelenterate Toxins*. Thesis, Harvard University, 1964.

Murtha, E.F. Pharmacological study of poisons from the shellfish and puffer fish. *Ann. N.Y. Acad. Sci*. 90, 820, 1960.

Murthy, S.S.N. and der Marderosian, A. In, *Food-Drugs from the Sea Proceedings 1972*. Worthen, L.R. (ed.), p. 181, Marine Technology Society: Washington, D.C., 1973.

Music, S.I., Howell, J.T. and Brumback, C.L. Red tide: its public health implications. *J. Fla. med. Assoc*. 60, 27, 1973.

Meyer, K.F., Sommer, H. and Schoenholz, P. Muscle poison. *J. prev. Med*. 2, 365, 1928.

Mynderse, J.S. and Moore, R.E. Toxins from blue-green algae: structures of oscillatoxin A and three related bromine-containing toxins. *J. Org. Chem*. 43, 2301, 1978.

Mynderse, J.S., Moore, R.E., Kashiwagi, M. *et al*. Antileukemic activity in the Oscillatoriaceae: isolation of debromoaplysiatoxin from *Lyngbya*. *Science* 196, 538, 1977.

Nagabhushanam, R., Kumar, A., Sarojini, R. *et al*. Toxicity of holothurian toxin in the commerical shrimp. *Toxicon*, 27, 65, 1989.

Nakagawa, H., Tu, A.T. and Kimura, A. Purification and characterization of contractin A from the pedicellarial venom of sea urchin, *Toxopneustes pileolus*. *Arch. Biochem. Biophys*. 284(2), 279, 1991.

Nallathambi, T., Kasinathan, R., Ravichandran, V. *et al*. Pharmacological actions of the venom of *Conus inscoiptus*. *10th World Congress on Animal, Plant and Microbial Toxins*. Singapore, Nov. 1991. Abst. 218, p. 270.

Naqvi, S.W.A., Kamal, S.Y., Solimabi, L. *et al*. Screening of some marine plants from the Indian coast for biological activity. *Botanica Marina* 24, 51, 1981.

National Marine Fisheries Service. *Oxygen Depletion and Associated Environmental Disturbances in the Middle Atlantic Bight in 1976*. N.M.I.T., N.E.F.C., Sandy Hook Laboratory, Highland, N.J., Tech. Ser. Rept. No. 3, 1, 1977.

Natori, T. [Ainu and Archaeology]. Vol. I, 202: Hokkaido Shuppan Kikaku Senlae: Sapporo. 1972.

Nayar, K.N. *Present Status of Molluscan Fisheries and Culture of India*. CMFR Institute: Cochin, India, 1980.

Neéman, I. Calton, G.J. and Burnett, J.W. Purification and characterization of the endonuclease present in *Physalia physalis* venom. *Comp. Biochem. Physiol*. 67B, 155, 1980a.

Neéman, I. Calton, G.J. and Burnett, J.W. An ultrastructural study of the cytotoxic effect of the venoms from the sea nettle (*Chrysaora quinquecirrha*) and Portuguese man-of-war (*Physalia physalis*) on cultured Chinese hamster ovary K-1 cells. *Toxicon* 18, 495, 1980b.

Neéman, I. Calton, J.G. and Burnett, J.W. Purification of an endonuclease present in *Chrysaora quinquecirrha* venom. *Proc. Soc. exp. Biol. Med.* 166, 374, 1981.

Neéman, I., Fishelson, L. and Kashman, Y. Sarcophine-a new toxin from the soft coral *Sarcophyton glaucum* (Alcyonaria). *Toxicon* 12, 593, 1974.

Newhouse, M.L. Dogger Bank itch: survey of trawlermen. *Br. med. J.* 1, 1142, 1966.

Nigrelli, R.F. The effects of holothurin on fish and mice with sarcoma 180. *Zoologica* 37, 89, 1952.

Nigrelli, R.F. and Jakowska, S. Effects of holothurin, a steroid saponin from the Bahamian sea cucumber (*Actinopyga agassizi*) on various biological systems. *Ann. N.Y. Acad. Sci.* 90, 884, 1960.

Nightingale, W.H. *Red Water Organisms: Their Occurrence and Influence Upon Marine Aquatic Animals, with Special Reference to Shellfish in Waters of the Pacific Coast.* Argus Press: Seattle, Washington, 1936.

Nitta, S. Über Nereistoxin, einen giftigen Bestandteil von *Lumbriconereis heteropoda* (Eunicidae). *J. Pharm. Soc. Jpn.* 54, 648, 1934.

Noguchi, T., Daigo, K., Arakawa, O. *et al.* Release of paralytic shellfish poison from the exoskeleton of a xanthid crab *Zosimus aeneus*. In, *Toxic Dinoflagellates.* Anderson, D.M., White, A.W. and Baden, D.G. (eds.), p. 293, Elsevier: New York, 1985.

Norton, R.S., Bobrek, G., Ivanov, I.D. *et al.* Purification and characterization of proteins with cardiac stimulatory and hemolytic activity from the anemone *Actinia tenebrosa*. *Toxicon* 28(1), 29, 1990.

Norton, T.R. Cardiotonic polypeptides from *Anthopleura xanthogrammica* (Brandt) and *A. Eligantissima* (Brandt). *Fed. Proc.* 40, 21, 1981.

Norton, T.R., Shibata, S., Kashiwagi, M. *et. al.* Isolation and characterization of the cardiotonic polypeptide Anthopleurin-A from the sea anemone *Anthopleura xanthogrammica*. *J. pharm. Sci.* 65, 1368, 1976.

Novak, V., Sket, D., Cankar, G. *et al.* Partial purification of a toxin from the tentacles of the sea anemone *Anemonia sulcata*. *Toxicon* 11, 411, 1973.

O'Connell, M.G. *Fine Structure of Venom Gland Cells in Globiferous Pedicellariae from Sea Urchins.* Thesis, Cal. State Coll.: Long Beach, 1971.

Olney, J.W., Teitelbaum, J., Pinsky, C. *et al.* Domoic acid toxicity. *Can. Dis. Wkly. Rept.* 16 Suppl. 1E, 117, 1990.

Orizumi, Y., Kobayashi, M., Kajiwara, A. *et al.* Potent excitatory effects of maitotoxin on cardiac and smooth muscle. In, *Toxic Dinoflagellates.* Anderson, D.M., White, A.W. and Baden, D.G. (eds.), p. 369, Elsevier: New York, 1985.

Othman, I., and Burnett, J.W. Techniques applicable for purifying *Chironex fleckeri* (box jellyfish) venom. *Toxicon* 28(7), 821, 1990.

Owellen, R.J., Owellen, R.G., Gorog, M.A. and Klein, D. Cytolytic fraction from *Asterias vulgaris*. *Toxicon* 11, 319, 1973.

Padilla, G.M. and Martin, D.F. Interactions of *Prymnesium parvum* toxin with erythrocyte membranes. In, *Tier und Pflanzengifte/Animal and Plant Toxins.* Kaiser E. (ed.), W. Goldmann, Munich: see also (1973) *Marine Pharmacognosy. Action of Marine Biotoxins at the Cellular Level.*

Martin, D.R. and Padilla M.S. (eds.), p. 265, Academic Press: New York, 1972; see also *Toxicon* 10, 532, 1972.

Pantin, C.F. The excitation of nematocysts. *J. exp. Biol.* 19, 294, 1942.

Paradice, W.E.J. Injuries and lesions caused by the bites of animals and insects. *Med. J. Aust.* 2, 650, 1924.

Parker, C.A. Poisonous qualities of the star-fish. *Zoologist* 5, 214, 1881.

Parker, G.H. and Van Alstyne, M.A. The control and discharge of nematocysts, especially in *Metridium* and *Physalia*. *J. exp. Zool.* 63, 329, 1932.

Parnas, I. and Abbott, B.C. Physiological activity of the ichthyotoxin from *Prymnesium parvum*. *Toxicon* 3, 133, 1965.

Parnas, I. and Russell, F.E. Effects of venoms on nerve, muscle and neuromuscular junction. In, *Animal Toxins*. Russell, F.E. and Saunders, P.R. (eds.), Pergamon Press: Oxford, 1967.

Parulekar, A.H. Cnidae in the actinians of Maharashtra. *J. Biol. Sci.* 9(1/2), 36, 1966.

Paster, Z. Purification and properties of prymnesin, the toxin formed by *Prymnesium parvum* (Chrysomonadinae). Ph.D. thesis. Hebrew University, Jerusalem, 1968.

Paster, Z. Pharmacognosy and mode of action of *Prymnesium*. In, *Marine Pharmocognosy. Action of Marine Biotoxins at the Cellular Level*. Martin, D.I. and Padilla, G.M. (eds.), p. 241, Academic Press: New York, 1973.

Paster, Z. and Abbott, B.C. Hemolysis of rabbit erythrocytes by *Gymnodinium breve* toxin. *Toxicon* 7, 245, 1969.

Pearn, J. *Animal Toxins and Man*. 123 pp. Courier Mail Printing Service: Brisbane, 1981.

Pawlowsky, E.N. *Gifttiere und Ihre Giftigkeit*. 516 pp. G. Fisher: Jena, 1927.

Pearn, J. *Animal Toxins and Man*. Queensland Health Department: Brisbane, 1981.

Peel, N. and Kandler, R. Localized neuropathy following jellyfish sting. *Postgrad Med. J.* 66, 953, 1990.

Pencar, S., Ferlan, I., Cotic, L. *et al*. An EPR study of the hemolytic action of equinatoxin. *Period. Biol.* 77, 149, 1975.

Penner, L.R. Bristleworm stinging in a natural environment. *Univ. Conn. Occ. Pap.* 1, 275, 1970.

Pennington, M.W., Kem, W.R., Norton, R.S. *et al*. Chemical synthesis of a neurotoxic polypeptide from the sea anemone *Stichodactyla helianthus*. *Intern. J. Peptide Prot. Res.* 36, 335, 1990.

Péréz, J.M. Recherches sur les pédicellaires glandulaires de *Sphaerechinus granularis* (Lamarck). *Arch. Zool. exp. gen.* 86, 118, 1950.

Perl, T.M., Bédard, L., Kosatsky, T. *et al*. An outbreak of toxic encephalopathy caused by eating mussels contaminated with domoic acid. *New England J. Med.* 322(25), 1775, 1990.

Phillips, C. and Brady, W.H. *Sea Pests: Poisonous or Harmful Sea Life of Florida and the West Indies*. Univ. Miami Press: Miami, 1953.

Phillips, J.H. Isolation of active nematocysts of *Metridium senile* and their chemical composition. *Nature* 178, 1932, 1956.

Phisalix, M. *Animaux Venimeux et Venins*. Masson et Cie.: Paris 1922.

Plumert, A. Über giftige Seetiere im Allgemeinen und einen Fall von Massenvergiftung durch Seemuscheln im Besonderen. *Arch. Schiffs-u. Tropenhyg.* 6, 15, 1902.

Poli, M.A., Templeton, C.B. and Thompson, W.L. *et al.* Distribution and elimination of brevetoxin PbTx-3 in rats. *Toxicon* 28, 903, 1990.

Pope, E.C. Sea animals that bite and sting. *Aust. Mus. Mag.* 9, 164, 1947.

Posner, P. and Kem. W.R. Cardiac effects of toxin A-III from the heteronemertine worm *Cerebratulus lacteus* (Leidy). *Toxicon* 16, 343, 1978.

Prakash, A., Medcof, J.C. and Tennant A.D. Paralytic shellfish poisoning in Eastern Canada. *Bull. Fish. Res. Bd. Canada* 117, 1, 1971.

Prasad, A.V.K. and Shimizu, Y. The structure of hemibrevetoxin-B: a new type of toxin in the Gulf of Mexico red tide organism. *J. Am. Chem. Soc.* 11, 6476, 1989.

Price, D.W. and Kizer, K.W. *California's Paralytic Shellfish Poisoning Program* 216, 1968.

Prinzmetal, M., Sommer, H. and Leake, C.D. The pharmacological action of "mussel poison". *J. Pharmacol. exp. Therap.* 46, 63, 1932.

Proctor, N.H., Chan, S.L. and Taylor, A.J. Production of saxitoxin by cultures of *Gonyaulax catenella*. *Toxicon* 13, 1, 1975.

Quilliam, M.A. and Wright, J.L.C. The amnesic shellfish poisoning mystery. *Analyt. Chem.* 61(18), 1053A, 1989.

Quinn, R.J., Gregson, R.P., Cook, A.F. *et al.* Isolation and synthesis of I-methy I-isoguanosine, a potent pharmacologically active constituent from the marine sponge *Tedania digitata*. *Tetrhed Lett.* 21, 567, 1980.

Rabbani, M.M. and Nisa, M. 1990. Mortality of Fish. *Sci. People; Tech. Progress*, p. 64.

Ragelis, E.P. *Seafood Toxins*. Am. Chem. Soc.: Washington, D.C., 1984.

Raj, U., Haq, Y., Oshima, Y. *et al.* The occurrence of paralytic shellfish toxins in two species of xanthid crab from Suva barrier reefs, Fiji Islands. *Toxicon* 21, 547, 1983.

Rajendran, N. and Kasinathan, R. A preliminary report of cone toxins of fishes and crabs. *Curr. Sci.* 56(22), 1176, 1987.

Randall, J. A review of ciguatera, tropical fish poisoning, with a tentative explanation of its cause. *Bull. mar. Sci. Gulf Caribb.* 8, 236, 1958.

Rao, D.S., Girijavallabhan, S., Muthusamy, V. *et al.* Bioactivity in marine algae. In, *Bioactive Compounds from Marine Organisms*. Thompson, M.F., Sarojini, R. and Nagbhushanam, R. (eds.), Oxford & IBH Publishing Co.: New Delhi, 1991.

Rao, D.S., James, D.B., Pillai, C.S.G. et al. Biotoxicity in marine organisms. In, *Bioactive Compounds from Marine Organisms*. Thompson, M.F., Sarojini, R. and Nagabhushanam, R. (eds.), Oxford & IBH Publishing Co.: New Delhi, 1991.

Rathmeyer, W., Jessen, B. and Beress, L. Effect of toxins of sea anemones on neuromuscular transmission. *Naturwissenschaften* 62, 538, 1975.

Ray, S.M. and Rao, K.S. *Manual on Marine Toxins in Bivalve Molluscs and General Consideration of Shellfish Sanitation*. Central Marine Fisheries Research Institute: Cochin, India, 1984.

Ray, S.M.R. *Manual of Marine Toxins in Bivalve Molluscs and General Consideration of Shellfish Sanitation*. C.M.F.R.I. Special Publ. #16 Cochin, India p. 100, 1984.

Read. B.E. Chinese materia medica. Fish drugs. *Peking Nat. Hist. Bull.* 136 pp., 1939.

Reich, K. and Kahn, J. A bacteria-free culture of *Prymnesium parvum* (Chrysomonadinae). *Bull. Res. Counc. Israel* 4, 144, 1954.

Reid, H.A. Symptomatology, pathology and treatment of the bites of sea snakes. In, *Snake Venoms*, Lee, C.-Y. (ed.), *Handbk, exp. Pharmacol.* Vol 52, p. 922, Springer Verlag: Berlin, 1979.

Rice, N.E. and Powell, W.A. Observations on three species of jelly-fishes from the Chesapeake Bay with special reference to their toxins. I. *Chrysaora (Dactylometra) quinquecirrha. Biol. Bull.* 139, 180, 1970.

Rich, A.C. Crayfish poisoning: a series of cases. *Liverpool Med. Chir. J.* 2, 384, 1882.

Richet, C. Anaphylaxie par la mytilo-congestine. *C.R. Cos. Biol.* 62, 358, 1907.

Rio G.J., Ruggieri, G.D., Stempein, M.F., Jr. and Nigrelli, R.F. Saponin-like toxin from the giant sunburst starfish *Pycnopodia helianthoides* from the Pacific Northwest. *Am. Zool.* 3, 544, 1963.

Roaf, H.E. and Nierenstein, M. Adrénaline et purpurine (reply to Dubois, R.). *C.R. Soc. Biol.* 63, 773, 1907.

Robson, E.A. The behaviour and neuromuscular system of *Gonacinia prolifera*, a swimming sea anemone. *J. Exp. Biol.* 55, 611, 1972.

Roughley, T.C. Where nature runs riot. *Nat. Geog.* 77(6) 823, 1940.

Ruggieri, G.D., Nigrelli, R.F. and Stempien, R.F., Jr. *et al.* Some biochemical and physiological properties of extracts from several echinoderms. *2nd Int. Symp. Animal Toxins,* p. 75, Tel Aviv, 1970.

Russell, F.S. *The Medusae of the British Isles.* Cambridge University Press: Cambridge, England, 1953.

Russell, F.E. Marine toxins and venomous and poisonous marine animals. In, *Advances in Marine Biology,* Russell, F.S. (ed.), Vol. 3, p. 255 Academic Press: London and New York, 1965(a).

Russell, F.E. Venomous and poisonous marine animals and their toxins. *First Inter-American Naval Res. Conf.,* San Juan, Puerto Rico, 37 pp., 1965(b).

Russell, F.E. To be, or not to be. . . the LD_{50}. *Toxicon* 4, 81, 83, 1966(a).

Russell, F.E. *Physalia* stings: a report of two cases. *Toxicon* 4, 65, 1966(b).

Russell, F.E. Poisonous marine animals. In, *Marine Toxins and Venomous and Poisonous Marine Animals.* T.F.H. Publications: Neptune City, N.J. 1971a.

Russell, F.E. Pharmacology of toxins of marine origin. In, *International Encyclopedia of Pharmacology and Therapeutics.* Raskova, H. (ed.) Sect. 71, vol. II, p.3, Pergamon Press: Oxford, 1971.

Russell, F.E. *Snake Venom Poisoning.* Lippincott: Philadelphia, 1980b. 552 pp.

Russell, F.E. Jellyfish stings. In, *Current Therapy.* Conn, H.F. (ed.) W.B. Saunders: Philadelphia, 1976.

Russell, F.E. Pharmacology of venoms. In, *Natural Toxins* (Proc. 6th Intern. Symp. Animal, Plant, Microb. Toxins, Uppsala 1979), Eaker D. and Wadström, T. (eds.), p. 13, Pergamon Press: Oxford, 1980(a).

Russell, F.E. Marine toxins and venomous and poisonous marine plants and animals (invertebrates.) In, *Advances in Marine Biology.* Russell, F.S. (ed.), Vol. 21, p. 59 Academic Press: London and New York (1984).

Saitanu, K., Wisessang, S., Piyakarnchana, T. *et al.* Occurrence of paralytic shellfish toxins and tetrodotoxin in green mussel (*Perna viridis*) and Sand clam (*Asaphis violescens*) which are not associated with toxic dinoflagellates. Proc. *10th Intern. Symp.* p. 291, 1991.

Salkowski, E. Zur Kenntniss des Giftes der Miesmuschel (*Mytilus edulis*). *Virchows Arch.* 102, 578, 1885.

Sarma, N.S., Rao, K.S. and Viswanadham, B. Settling responses and progression in community development of selected macrofouling organisms to a recently isolated sponge metabolite, herbacin, at Visakhapatanam Harbor, Bay of Bengal. In, *Bioactive Compounds from Marine Organisms*. Thompson, M.F., Sarojini, R. and Nagabhushanam, R. (eds.), Oxford & IBH Publishing Co.: New Delhi, 1991.

Savage, I.V.E. and Howden, M.E.H. Hapalotoxin, a second lethal toxin from the octopus *Hapalochlaena maculosa*. *Toxicon* 15, 463, 1977.

Saville-Kent, W. *The Great Barrier Reef of Australia — Its Products and Potentialities.* W.H.Allen & Co.: London, 1893.

Saunders, R.D. and Dodge, J.D. An SEM study and taxonomic revision of some amoured sand-dwelling dinoflagellates. *Protistologica* 20, 271, 1984.

Schantz, E.J. Biochemical studies on paralytic shellfish poisons. *Ann. N.Y. Acad. Sci.* 90, 843, 1960.

Schantz, E.J. Studies on the paralytic poisons found in mussels and clams along the North American Pacific Coast. In, *Venomous and Poisonous Animals and Noxious Plants of the Pacific Region*. Keegan, H.H. and MacFarlane, M.V. (eds.), p. 75, Pergamon Press: Oxford, 1963.

Schantz, E.J. The dinoflagellate poisons. In, *Microbial Toxins. Algal and Fungal Toxins*. Kadis, S., Ciegler, A. and Ajkedsi, S. (eds.), Vol. III, p. 3, Academic Press: New York, 1971.

Schantz, E.J., Lynch, J.M., Vayvada, G. et al. The Purification and characterization of the poison produced by *Gonyaulax catenella* in axenic culture. *Biochemistry* 5, 1191, 1966.

Schantz, E.J. Ghazarossian, V.E., Schnoes, H.K. *et al.* Paralytic poisons from marine dinoflagellates. In, *Proceedings of the First International Conference on Toxic Dinoflagellate Blooms*. Locicero, V.R. (ed.), p. 267, Mass. Sci. Tech. Found.: Wakefield, Massachusetts, 1975.

Schantz, E.J., Lynch, J.M. Vayvada, G. *et al.* The purification and characterization of the poison produced by *Gonyaulax catenella* in axenic culture. *Biochemistry* 5, 1191, 1966.

Schantz, E.J., McFarren, E.F., Schaffer, M.L. *et al.* Purified shellfish poison for bioassay standardization. *J. Assoc. Off. Agric. Chem.* 41, 160, 1958.

Schantz, E.J., Mold, J.D., Stanger, D.W. *et al.* Paralytic shellfish poison VI. A procedure for the isolation and purification of the poison from toxic clam and mussel tissues. *J. Am. chem. Soc.* 79, 5230, 1957.

Scheuer, P.J. *Chemistry of Marine Natural Products.* Academic Press: New York and London, 1973.

Scheuer, P.J. Recent developments in the chemistry of marine toxins. *Lloydia* 38, 1, 1975.

Scheuer, P.J. Marine toxins. *Acc. Chem. Res.* 10, 33, 1977.

Scheuer, P.J. *Marine Natural Products.* Vol. II, Academic Press: London, 1978.

Scheumack, D.D., Howden, M.E.H., Spence, I. *et al.* Maculotoxin: a neurotoxin from the venom glands of octopus *Hapalochlaena maculosa* identified as tetrodotoxin. *Science* 199, 188, 1978.

Schmitz, F.J., Gunasekera, S.R., Yalamanchili, G. *et al.* Tedanolide: a potent cytotoxic macrolide from the Caribbean sponge *Tedania ignis*. *J. Am. Chem. Soc.* 106, 7251, 1984.

Schmitz, F.J., Hollenbeak, K.H. and Campbell, D.C. Marine natural products: halitoxin, toxic complex of several marine sponges of the genus *Haliclona*. *J. org. Chem.* 43, 3916, 1978.

Schulte, B.A. and Bakus, G.J. Predation deterrence in marine sponges: laboratory versus field studies. *Bull. Mar. Sci.* 50, 205, 1992.

Schweitz, H., Vincent, J.P., Baharnin, J. *et al.* Purification and pharmacological properties of eight sea anemone toxins from *Anemonia sulcata. Anthopleura xanthogrammica. Stoichactis giganteus* and *Actinodendron plumosum. Biochemistry* 20, 5245, 1985.

Schweitz, H., Kallenbach, N.R., Wemmer, D.E. NMR studies of toxin III from the sea anemone *Radianthus paumotensis* and comparison of the secondary structure with related toxins. *Biochem.* 28, 2199, 1989.

Sencec, L. and Macek, P. New method for isolation of venom from the sea anemone *Actinia cari*. Purification and characterization of cytolytic toxins. *Comp. Biochem. Physiol.* 97B(4), 687, 1990.

Seville R.H. Dogger Bank itch. Report of a case. *Br. J. Derm.* 69, 92, 1957.

Shanmuganandam, P., Kasinathan, R., Babuji, S. *et al.* Pharmacological screening of *Conus virgo* venom on cardiac musculature. *10th World Congress on Animal Plant and Microbial Toxins*, Singapore, Nov. 1991. Abst. 240.

Shapiro, B.I. Purification of a toxin from the tentacles of the anemone *Condylactis gigantea. Toxicon* 5, 253, 1968a.

Shapiro, B.I. A site of action of toxin from the anemone *Condylactis gigantea. Comp. Biochem. Physiol.* 27A,B,C, 519, 1968b.

Shapiro, B. I. and Lilleheil, G. The action of anemone toxin on crustacean neurons. *Comp. Biochem. Physiol.* 28A,B,C, 1225, 1969.

Sheridan, R.E. and Adler, M. The actions of a red tide toxin from *Ptychodiscus brevis* on single sodium channels in mammalian neuroblastoma cells. *F.E.B.* 247, 448, 1989.

Shibota, M. and Hashimoto, Y. Purification of the ivory shell toxin. *Bull. Jpn. Soc. scient. Fish* 36, 115, 1970.

Shilo, M. and Rosenberger, R.F. Studies on the toxic principles formed by the chrysomonad *Prymnesium parvum* Carter. *Ann. N.Y. Acad. Sci.* 90, 866, 1960.

Shimada, S. Antifungal steroid glycoside from sea cucumber. *Science* 163, 1462, 1969.

Shimizu, Y. Dinoflagellate toxins. In, *Marine Natural Products*. Scheuer, P.J. (ed.), Vol. 1, p. 1. Academic Press: New York, 1978.

Shimizu, Y. Dinoflagellate toxins. In, *The Biology of Dinoflagellates*. p. 282, Taylor, F.J.R., (ed.), Blackwell: Oxford, 1978.

Shimizu, Y. Chemistry of paralytic shellfish poisons. In, *Progress in the Chemistry of Organic Natural Products*. Herz, U., Grisebach, H. and Kirby, G.W. (eds.), p. 235, Springer: New York, 1984.

Shimizu, Y., Alam, M., Oshima, Y. *et al.* Presence of four toxins in red tide infested clams and cultured *Gonyaulax tamarensis* cells. *Biochem. Biophys. Res. Commun.* 66, 731, 1975

Shimizu, Y. and Hsu, C.P. Confirmation of the structures of Gonyaulax I-IV by correlation with saxitoxin. *Chem. Commun.* 314, 1981.

Shimizu, Y., Hsu, C.P. and Genenah, A. Structure of saxitoxin in solution and stereochemistry of dihydrosaxitoxin. *J. A. Chem. Soc.* 103, 605, 1981.

Shimizu, Y., Kobayashi, M. Genenah, A. *et al.* Biosynthesis of paralytic shellfish toxins. In, *Seafood Toxins*. Ragelis, E. (ed.), ACS Symposium Series: Washington, D.C., 1984.

Shimizu, Y., Painuli, P., Gupta, S. *et al.* Dinoflagellate and other algal toxins: chemistry and toxicology. In, *Bioactive Compounds from Marine Organisms*. Thompson, M.F., Sarojini, R. and Nagabhushanam, R. (eds.), Oxford & IBH Publishing Co.: New Delhi, 1991.

Shin, L.M., Michaels, W.D. and Mayer, M.M. Membrane damage by a toxin from the sea anemone *Stoichactis helianthus*-II Effect of membrane lipid composition in a liposome system. *Biochim. Biophys. Acta* 555, 79, 1979.

Shiomi, K., Yamamoto, S., Yamanaka, H. *et al.* Liver damage by the crown-of-thorns starfish (*Acanthaster planci*) lethal factor. *Toxicon* 28(5), 469, 1990.

Shumway, S.E. A review of the effects of algal blooms on shellfish and aquaculture. *J. World. Aquacult. Soc.* 21, 65, 1990

Silas, E.G., Algariswami, K.A., Narasimham, K.K. *et al.*, 1982 Country Report—India. In, *Bivalve Culture in Asia and the Pacific*. Proc. Workshop, Singapore. Ottawa, Ont. IDRC, p. 34, 16, 34, Feb. 1982.

Simon, S.E., Cairncross, K.D., Stachell, D.G. et al. The toxicity of *Octopus maculosus* Hoyle venom. *Acrh. intern. Pharmacodym.* 148, 318, 1964.

Sims, J.K. and Irei, M.Y. Human Hawaiian marine sponge poisoning. *Hawaii med. J.* 38, 263, 1979.

Sims, J.K. and Zandee van Rilland, R.D. Escharotic stomatitis caused by the "stinging seaweed" *Microcoleus lyngbyaceus*. *Hawaii med. J.* 40, 243, 1981.

Sket, D., Draslar, K., Ferlan, I. *et al.* Equinatoxin, a lethal protein from *Actinia equina*-II. Pathophysiological action. *Toxicon* 12, 63, 1974.

Smith, F.G.W. Red tide studies. *Mar. Lab. Univ. Miami, Coral Gables* (Prelim. Rep. to Fla. St. Bd. Conservation, 1954.

Snodgrass, S.R. Neurologic sequelae after ingestion of mussels contaminated with domoic acid. *New Engl. J. Med.* 323(23), 1631, 1990.

Snow, C.D. Two accounts of the southern octopus, *Octopus dofleini* biting scuba divers. *Oregon Fish. Comm. Res. Rep.* 2, 103, 1970.

Solomon, A.E. and Stoughton, R.B. Dermatitis from purified sea algae toxin (debromoaplysiatoxin) *Arch. Dermatol.* 114, 1333, 1978.

Sommer, H. The occurrence of the paralytic shell-fish poison in the common sand-crab. *Science* 76, 574, 1932.

Sommer, H. and Meyer, K.F. Paralytic shellfish poisoning. *Arch. Path.* 24, 560, 1937.

Sommer, H., Monnier, R.P., Riegel, B. *et al.* Paralytic shellfish poison. I. Occurrence and concentration by ion exchange. *J. Am. chem. Soc.* 70, 1015, 1948.

Southcott, R.V. Coelenterates of medical importance. In: *Venomous and Poisonous Animals and Noxious Plants of the Pacific Region*. Keegan, H.S. and MacFarlane, W.P. (eds.), Pergamon Press: New York, 1963.

Southcott, R.V. *Australian Venomous and Poisonous Fishes*. Southcott: Mitchim, South Australia, 1975.

Southcott, R.V. and Coulter, J.R. The effects of southern Australian marine stinging sponges, *Neofibularia mordens* and *Lissodendoryx* sp. *Med. J. Aust.* (2), 895, 1971.

Southcott, R.V. and Kingston, C.W. Lethal jellyfish stings: a study in sea wasps. *Med. J. Aust.* 1, 443, 1959.

Spector, I., Shochet, N.R., Kashman, Y. *et al.* Latrunculines: novel marine toxins that disrupt microfilament organization in cultured cells. *Science* 219, 493, 1983.

Spense, I. Long-lasting, reversible spastic paresis produced by a halogenated monoterpene isolated from the red alga *Plocanium carilagineum* (L.). 7th *Internat. Congr. Pharmacol.*, Abst 84, 1978.

Spense, I., Jamieson, D.D. and Taylor, K.M. Anticonvulsant activity of farnesylacetone epoxide—a novel marine product. *Experientia* 35, 238, 1979.

Starr, T.J. Notes on a toxin from *Gymnodinium brevis*. *Texas Rept. Biol. Med.* 20, 271, 1958.

Steidinger, K.A. Basic factors influencing red tides. In, *Toxic Dinoflagellate Blooms*. Locicero, V.R. (ed.), p. 152, Mass. Sci. Tech. Found.: Wakefield, MA, 1975.

Steidinger, K.A. A re-evaluation of toxic dinoflagellate biology and ecology. In, *Progress in Phycological Research*. Round, F.E. and Chapman, D.J. (eds.), Elsevier: New York, 1983.

Steidinger, K.A. Some taxonomic and biologic aspects of toxic dinoflagellates. In, *Algal Toxins in Seafood and Drinking Water*. Falconer, I.R. and Hald, B. (eds.), Academic Press: London and New York, 1993.

Steidinger, K.A., Burklew, M.A. and Ingle, R.M. Effects of *Gymnodinium breve* toxin on esturaine animals. In, *Marine Pharmacognosy*. Martin, D.F. and Padilla, G.M. (eds.), p. 127, Academic Press: New York, 1973.

Stein, M.R., Marracini, J.V., Rothschild, N.E. *et al.* Fatal Portuguese Man-O' War (*Physalia physalis*) envenomation. *Ann. Emerg. Med.* 18, 312, 1989.

Stempien, M.F., Ruggieri, G.D., Nigrelli, R.F. et al. Physiologically active substances from extracts of marine sponges. In, *Food-Drugs from the Sea Conference*. Youngken, H.W. (ed.), p. 295, *Mar. Tech. Soc.*: Washington, D.C., 1970.

Stillway, L.W. and Lane, C.E. Phospholipase in the nematocyst toxin of *Physalia physalis*. *Toxicon* 9, 193, 1971.

Stonik, V.A. and Elyakov, G.B. Structure and biologic activities of sponge and sea cucumber toxins. In, *Handbook of Natural Toxins*. Tu, A.T. (ed.), vol. III, p. 107, Marcel Dekker: New York, 1988.

Strazisar, M., Sedmak, B., Kotnik, V. *et al.* The cellular immune response to Eq T II, a cytolytic and lethal toxin from the sea anemone *Actinia equina*. *Toxicon* 27(1), 80, 1989.

Stuart-Harris, C. The Garrod lecture 1984. Strategies of anti-viral chemotherapy. *J. Antimicrob. Chemother.* 15, 387, 1985.

Subrahmanyan, R. On the life history and ecology of *Hornelia marina* (gen et sp. nov. Chloromonadineae) caused green discoloration of the sea and mortality among marine organisms off the Malabar Coast. *India J. Fish.* 1, 182, 1954.

Subrahmanyan, R. Phytoplankton organisms of the Arabian Sea of the West Coast of India. *J. Indian Bot. Soc.* 37, 435, 1958.

Subrahmanyan, A. Noxious dinoflagellates in Indian waters. In, *Toxic Dinoflagellates*. Anderson, D.M., White, A.W. and Baden, D.G. (eds.), p. 525, Elsevier: New York, 1985.

Sullivan, J.J., Jonas-Davies, B. and Kentala, L.L. The determination of PSP toxins by HPLC and autoanalyzer. In, *Toxic Dinoflagellates*. Anderson D.M., White, A.W. and Baden, D.G. (eds.), p. 275, Elsevier: New York, 1985.

Sutherland, S. *Australian Animal Toxins: The Creatures, Their Toxins and Care of the Poisoned Patient.* Oxford University Press: Melbourne, 1983.

Sutherland, S.K., Broad, A.J. and Lane, W.R. Octopus neurotoxins: low molecular weight non immunogenic toxins present in the saliva of the blue-ringed octopus. *Toxicon* 8, 249, 1970.

Sutherland, S.K. and Lane, W.R. Toxins and mode of envenomation of the common ringed or blue-banded octopus. *Med. J. Aust.* 1, 893, 1969.

Sutherland, R.J., Hoesing, J.M. Whishaw, I.Q. Domoic acid, an environmental toxin, produces hippocampal damage and severe memory impairment. *Neurosci. Lett.* 120(2), 221, 1990.

Tachibana, K. *Structural Studies on Marine Toxins*. Dissertation, University of Hawaii, 1980.

Takei, M., Nakagawa, H., Kimura, A. *et al.* A toxic substance from the sea urchin *Toxopneustes pileolus* induces histamine release from rat peritoneal mast cells. *Agents and Actions* 32(3/4), 224, 1991.

Tamkun, M.M. and Hessinger, D.A. Isolation and partial characterization of a hemolytic and toxic protein from the nematocyst venon of the Portuguese man of war *Physalia physalis. Biochim. Biophys. Acta* 667, 87, 1981.

Tamiyanich, S., Kodama, M. and Fukuyo, Y. The occurrence of paralytic shellfish poisoning in Thailand. In, *Toxic Dinoflagellates*. Anderson, D.M., White, A.W. and Baden, D.G. (eds.), p. 521, Elsevier: New York, 1985.

Tanaka, M., Haniu, M., Yasunobu, K.T. et al. Amino acid sequence of the *Anthopleura xanthogrammica* heart stimulant, Anthopleurin A. *Biochemistry* 16, 204, 1977.

Taylor, F.J.R. Toxic Dinoflagellates: Taxonomic and biogeographic aspects with emphasis on *Protogonyaulax*. In. *Seafood Toxins*. Ragelis, E.P. (ed.), American Chemical Society: Washington, D.C., 1984.

Taylor, F.J.R. The taxonomy and relationships of red tide flagellates. In, *Toxic Dinoflagellates*. Anderson, D.M., White, A.W. and Baden, D.G. (eds.), Elsevier: New York, 1985.

Taylor, F.J.R. Red tides, brown tides and other harmful algae blooms: the view into the 1990's. In, *Toxic Marine Phytoplankton*, Graneli, E. (ed.), p. 527, Elsevier: Amsterdam, 1990.

Taylor, D.L. and Seliger, H.H. (eds.), *Toxic Dinoflagellate Blooms*. Elsevier/North-Holland: New York, 1979.

Taylor, K.M. and Spense, I. Marine natural products affecting neurotransmission. In, *Neurotoxins. Fundamental and Clinical Advances*. Chubb, I.W. and Getten L.B. (eds.), p. 85, Adelaide Univ. Union Press: Adelaide, S. Australia, 1979.

Teh, Y.F. and Gardiner, J.E. Partial purification of *Lophozozymus pictor* toxin. *Toxicon* 12, 603, 1974.

Teitelbaum, J.S., Zatorre, R.J., Carpenter, S. *et al.* Neurologic sequelae of domoic acid intoxication due to the ingestion of contaminated mussels. *New Engl. J. Med.* 322 (25), 1781, 1990.

Templeton, C.B., Polis, M.A. and LeClaire, R.D. Cardiorespiratory effects of brevetoxin (Pb-Tx-2) in conscious, tethered rats. *Toxicon* 27, 1043, 1989.

Teng, C-M., Lee, L-G., Lee, C-Y. *et al.* Platelet aggregation induced by equinatoxin. *Thromb. Res.* 52. 401, 1988.

Tennant, A.D., Naubert, J. and Corbeil, H.E. An outbreak of paralytic shellfish poisoning. *J. Canadian Med. Assoc.* 72, 436, 1955.

Thomas, P.A. Distribution and affinities of the sponge fauna of the Indian region. *J. Mar. Biol. Ass. India* 25, 7, 1983.

Thompson, J.E., Walker, R.P. and Faulkner, D.J. Screening and bioassay for biologically active substances from forty marine sponge species from San Diego, California, U.S.A. *Mar. Biol.* 88, 11, 1985.

Thompson, M.F., Sarojini, R. and Nagabhushanam, R. (eds.), *Bioactive Compounds from Marine Organisms.* Oxford & IBH Publishing Co.: New Delhi, 1991.

Thompson, T.E. Epidermal acid-secretion in some marine polyclad turbelliara. *Nature* 206, 954, 1965.

Thorington, G.U. and Hessinger, D.A. Control of cnida discharge: I. Evidence for two classes of chemoreceptor. *Biol. Bull.* 174, 163, 1988.

Thorington, G.U. and Hessinger, D.A. Control of cnida discharge: III. Spirocysts are regulated by three classes of chemoreceptor. *Biol. Bull.* 178, 74, 1990.

Thorpe, J.P. and Ryland, J.S. Cryptic speciation detected by biochemical genetics in three ecologically important intertidal bryozoans. *Estuarine Coast. Mar. Sci.* 8, 395, 1979.

Thron, C.D., Durant, R.C. and Friess, S.L. Neuromuscular and cytotoxic effect of holothurin A and related saponins at low concentration levels. III. *Toxicol. appl. Pharmacol.* 6, 182, 1964.

Thron, C.D., Patterson, R.N. and Friess, S.L. et al. Further biological properties of the sea cucumber toxin holothurin A. *Toxicol. appl. Pharmacol.* 5, 1, 1963.

Tindall, D.R. and Miller, D.M. Purification of maitotoxin from the dinoflagellate, *Gambierdiscus toxicus* using high pressure liquid chromatography. In, *Toxic Dinoflagellates.* Anderson, D.M., White, A.W. and Baden, D.G. (eds.), p. 321, Elsevier: New York, 1985.

Tindall, D.R., Miller, D.M. and Bomber, J.W. Culture and toxicity of dinoflagellates from ciguatera regions of the world. *Toxicon* 27, 83, 1989.

Todd, E.C.D. Ciguatera poisoning in Canada. In, *Toxic Dinoflagellates.* Anderson, D.M., White, A.W. and Baden, D.G. (eds.), p. 325, Elsevier: New York, 1985.

Todd, E.C.D. Amnesic shellfish poisoning—A new seafood toxin syndrome. In, *Toxic Marine Phytoplankton.* Graneli, E. (ed.), p. 504, Elsevier: Amsterdam, 1990.

Togashi, M. Clinical studies of the poisoning by *Venerupis semidecussata.* (In Japanese). *Jap. Iji Shimpo.* May, p. 31, 1943.

Tomlin, E.W.F. the Ainu: their history and culture. *J. Roy. Cent. Asia Soc.* 53, 297, 1966.

Trethewie, E.R. Pharmacological effects of the venom of the common octopus *Hapalochlaena maculosa. Toxicon* 3, 55, 1965.

Trevan, J.W. The error of determination of toxicity. *Proc. Roy. Soc.* (London) B101, 483, 1927.

Treiff, N.M., Venkatas, N. and Ray, S.M. Purification of *Gymnodinium breve* toxin—dry column chromatographic techniques. *Tex. J. Sci.* 23, 596, 1972.

Tsutsumi, J. and Hashimoto, Y. Isolation of pyopheophorbide a as a photodynamic pigment from the liver of abalone *Haliotis discus hannai. Agric. Biol. Chem.* 28, 467, 1964.

Turk, J.L., Parker, D. and Rudner, E.J. Preliminary results on the purification of the chemical sensitizing agent in *Alcyonidium gelatinosum. Proc. Roy. Soc. Med.* 59, 1122, 1966.

Turlapaty, P., Shibata, S., Norton, T.R. *et al.* A possible mechanism of action of a central stimulant substance isolated from sea anemone *Stoichactis kenti. Eur. J. Pharmac.* 24, 310, 1973.

Uda, I., Itoh, Y. and Nishimura, M. *et al.* Enzyme immunoassay using monoclonal antibodies specific for diarrhetic shellfish poisons. In, *Mycotoxins and Phycotoxins '88.* p. 335, Natori, K., Hashimoto, K. and Veno, E. (eds.), Elsevier: Amsterdam, 1989.

Uemura, D., Hirata, Y., Iwashita, T. *et al.* Studies on palytoxins. *Tetrahedron* 41, 1007, 1985.

Uemura, D.L., Ueda, K., Hirata, Y. *et al.* Structural studies on palytoxin, a potent coelenterate toxin. *Tetrahedron Lett.* 21, 4867, 1980.

Uemura, D., Ueda, K., Hirata, Y. *et al.* Further studies on palytoxin. I. *Tetrahedron Lett.* 22, 1909, 1981a.

Uemura, D., Ueda, K., Hirata, Y. *et al.* Further studies on palytoxin. II. *Tetrahedron Lett.* 22, 2781, 1981b.

Ulitzur, S. and Shilo, M. Effects of *Prymnesium parvum* toxins, acetyl trimethylammonium bromide and sodium dodecyl sulphate on bacteria. *J. gen. Microbiol.* 62, 363, 1970.

Vick, J.A. and Wiles J.S. The mechanism of action and treatment of palytoxin poisoning. *Toxicol. appl. Pharmacol.* 34, 214, 1975.

Von Uexkull, J. Die Physiologie der Pedicellarien. *Z. Biol. Ser.* 2.37, 344, 1899.

Walker, M.J.A. Coelenterate and echinoderm toxins: mechanisms and actions. In, *Handbook of Natural Toxins.* Tu, A.T. (ed.), vol. III, p. 279, Marcel Dekker: New York, 1988.

Walker, M.J.A. and Masuda, V.L. Toxins from marine invertebrates. *Am. Chem. Soc.* 418, 1990.

Wang, C.M., Narahashi, T. and Mendi, T.J. Depolarizing action of *Haliclona* toxin on end plate and muscle membranes. *Toxicon* 11, 499, 1973.

Waraszkiewicz, S.M. and Erickson, K.L. Halogenated sesquiterpenoids from the Hawaiian marine alga *Laurencia nidifica. Tetrahedron Lett.* 23, 2003, 1974.

Warnick, J.E., Weinreich, D. and Burnett, J.W. Sea nettle (*Chrysaora quinquecirrha*) toxin on electrogenic and chemosensitive properties of nerve and muscle. *Toxicon* 19, 361, 1981.

Watson, G.M. and Hessinger, D.A. Cnidocytes and adjacent supporting cells from receptor-effector complexes in *Anemonia* tentacles. *Tissue and Cell* 21, 17, 1989a.

Watson, G.M. and Hessinger, D.A. Cnidocyte mechanoreceptors are tuned to the movements of swimming prey by chemoreceptors. *Science* 243, 1589, 1989b.

Watson, G.M. and Hessinger, D.A. Chemoreceptor-mediated elongation of stereocilium bundles tunes vibration-sensitive mechanoreceptors on cnidocyte-supporting cell complexes to lower frequencies. *J. Cell. Sci.* 99, 307, 1991.

Watson, G.M. and Hessinger, D.A. Receptors for N-acetylated sugars may stimulate adenylate cyclase to sensitise and tune mechano-receptors involved in triggering nematocyst discharge. *Exp. Cell. Res.* 198, 8, 1992.

Watson, G.M. and Hessinger, D.A. Antagonistic frequency tuning of hair bundles by different chemoreceptors regulates nematocyst discharge. *J. exp. Biol.* 187, 57, 1994.

Watson, M. Midgut gland toxins of Hawaiian sea hares-I. Isolation and preliminary toxicology observations. *Toxicon* 11, 259, 1973.

Watson, M. and Rayner, M.D. Midgut gland toxins of Hawaiian sea hares-II. A preliminary pharmacological study. *Toxicon* 11, 269, 1973.

Weill, R. *Contributions à l'Etude des Cnidaires et de leurs Nématocystes.* Trav. Stat. Zool. Wimiereux: Paris, 1934.

Wells, R.J. New metabolites from Australian marine species. *Pure appl. Chem.* 51, 1829, 1955.

Wells, R.J. New metabolites from Australian marine species. *Pure appl. Chem.* 51, 1829, 1979.

Welsh, J.H. In: *The Biology of Hydra.* Lenhoff, H.M. and Loomis, W.F. (eds.), University of Miami, Miami, Florida, 1961.

Welsh, J.H. and Moorhead, M. The quantitative distribution of 5-hydroxytryptamine in the invertebrates, especially in their nervous system. *J. Neurochem.* 6, 146, 1960.

Werner, B. Die Nesselkapseln der Cnidaria mit besonderen Berücksichtigung der Hydroida. I Klassification und Bedeutung für die Systematik und Evolution. *Helgol. Wiss. Meeresunters.* 12, 1, 1965.

White, A.W. Dinoflagellate toxins as probable cause of an Atlantic herring (*Clupea harengus harengus*) kill, and pteropods as an apparent vector. *J. Fish. Res. Bd. Can.* 34, 2421, 1977.

Whittaker, V.P. Pharmacologically active choline esters in marine gastropods. *Ann. N.Y. Acad. Sci.* 90, 695, 1960.

Whittaker, V.P. and Wijesundera, S. The separation of esters of choline by filter-paper chromatography. *Biochem. J.* 51, 348, 1952.

White, D.R.L. and White, A.W. First report of paralytic shellfish poisoning in Newfoundland. In, *Toxic Dinoflagellates.* Anderson, D.M., White, A.W. and Baden, D.G. (eds.), p. 511, Elsevier: New York, 1985.

Whysner, J.A. and Saunders, P.R. purification of the lethal fraction of the venom of the marine snail *Conus californicus.* *Toxicon* 4, 177, 1966.

Whyte, J.M. and Endean, R. Pharmacological investigation of the venoms of the marine snails *Conus textile* and *Conus geographus.* *Toxicon* 1, 25, 1962.

Wiberg, G.A. and Stephenson, N.R. Toxicologic studies on paralytic shellfish poison. *Toxicol. Appl. Pharmacol.* 2, 607, 1960.

Wiles, J.S., Vick, J.A. and Christensen, M.K. Toxicological evaluation of palytoxins in several animal species. *Toxicon* 12, 427, 1974.

Williamson, J. *The Marine Stinger Book.* Queensland State Centre: Brisbane, 1985.

Winkler, L.R. Preliminary tests of the toxin extracted from California sea hares of the genus *Aplysia. Pac. Sci.* 15, 211, 1961.

Winkler, L.R., Tilton, B.E. and Hardinge, M.G. A cholinergic agent extracted from sea hares. *Arch. int. Pharmacodyn.* 137, 76, 1962.

Withers, N.W. Ciguatera research in the northwestern Hawaiian Islands. Laboratory and field studies on ciguatoxigenic dinoflagellates in the Hawaiian Archipelago. *Proc. Res. Inv. NWHI. UNIHI—Seagrant Ma 84-01*, 1984.

Wittle, L.W., Middlebrook, R.E. and Lane, C.E. Isolation and partial purification of a toxin from *Millepora alcicornis. Toxicon* 9, 327, 1971.

Wittle, L.W., Scura, E.D. and Middlebrook, R.E. Stinging coral (*Millepora tenera*) toxin: a comparison of crude extract with isolated nematocyst extracts. *Toxicon* 12, 481, 1974.

Wittle, L.W. and Wheeler, C.A. Toxic and immunological properties of stinging coral toxin. *Toxicon* 12, 487, 1974.

Wong, J.L., Oesterlin, R. and Rapoport, H. The structure of saxitoxin. *J. Am. Chem. Soc.* 93, 7344, 1971.

Woodward, G. *Conference on Shellfish Toxicology*, Public Health Service: Washington, D.C., 1955.

Yaffee, H.S. Irritation from red sponge (letter). *N. E. J. Med.* 282, 51, 1963.

Yaffee, H.S. and Stargardter, F. Erythema multiforme from *Tedania ignis. Arch. Derm.* 87, 601, 1963.

Yamamura, S. and Hirata, Y. Structures of aplysin and aplysinol, naturally occurring bromo-compounds. *Tetrahedron* 19, 1485, 1963.

Yamanouchi, T. Notes on the behavior of the holothurian *Caudina chilensis* (J. Muller). *Sci. Rep. Res. Inst. Tohoku Univ. (Biol.),* 4, 73, 1929.

Yamanouchi, T. Study of poisons contained in holothurians. *Teikoku Gakushiin Hokoku,* 17, 73, 1942.

Yamanouchi, T. On the poisonous substance contained in holothurians. *Publ. Seto Mar. Biol. Lab.* 4, 184, 1955.

Yanagita, T.M. Physiological mechanism of nematocyst responses in sea anemone II. Effects of electrolyte ions upon the isolated cnide. *J. Fac. Sci. Univ. Tokyo, Sect. IV* 8, 381, 1959a.

Yanagita, T.M. Physiological mechanism of nematocyst responses in sea anemone VII. Extrusion of resting cnidae — its nature and its possible bearing on the normal nettling response. *J. exp. Biol.* 36, 4178, 1959b.

Yanagita, T.M. The "cnidblast" as an excitable system. In, *Recent Trends in Research in Coelenterate Biology*. Tokioka, T. and Nishimura, S., (eds.), vol. 20, Publ. Seto Mar. Biol Lab.: Shirahama, Wakayama-Kem, Japan, 1973.

Yariv, J. and Hestrin, S. Toxicity of the extracellular phase of *Prymnesium parvum* cultures. *J. gen. Microbiol.* 24, 165, 1961.

Yasumoto, T. Recent progress in the chemistry of dinoflagellate toxins. In, *Toxic Dinoflagellates.* Anderson, M., White, A.W. and Baden, D.G. (eds.), p. 259, Elsevier: New York, 1985.

Yasumoto, T. and Hashimoto, Y. Properties and sugar components of asterosaponin A isolated from starfish. *Agric. Biol. Chem.* 31, 368, 1965.

Yasumoto, T., Nakamura, K. and Hashimoto, Y. A new saponin, holothurin B, isolated from sea cucumber, *Holothuria vagabunda* and *Holothuria lubrica. Agric. biol. Chem.* 31, 7, 1967.

Yasumoto, T., Tanaka, M., Hashimoto, Y. *et al.* Distribution of saponin in echinoderms. *Bull. Jpn. Soc. scient. Fish.* 32, 673, 1966.

Yasumoto, T., Watanabe, T. and Hashimoto, Y. (Physiological activities of starfish saponin.) *Bull. Jpn. Soc. Scient. Fish.* 30, 357, 1964.

Yasumoto, T., Yasumura, D., Ohizumi, Y. *et al.* Palytoxin in two species of xanthid crab from the Philippines. *Agric. Biol. Chem.* 50, 163, 1986.

Yentsch, C.M. Flow cytometric analysis of cellular saxitoxin in the dinoflagellate *Gonyaulax tamarensis* var. *excavata. Toxicon* 19, 611, 1981.

Zatorre, R.J. Memory loss following domoic acid intoxication from ingestion of toxic mussels. *Can. Dis. Wkly. Rept.* 16 Suppl. 1E, 1 1989.

Fisher, A.A. *Atlas of Aquatic Dermatology*. Grune and Stratton, New York, 1978.

Flecker, H. and Cotton, B.C. Fatal bite from octopus. *Med. J. Aust.* (2), 329, 1955.

Fleming, W.J. *et al.* Partial purification and characterization of a lethal protein from *Tripneustes gratilla*. *Toxicon*, 12, 447, 1974.

Fleming, W.J., Salathe, R., Wyllie S.G. and Howden, M.E.H. Isolation and partial characterization of steroid glycosides from the starfish *Acanthaster planci*. *Comp. Biochem. Physiol.* 53B. 267, 1976.

Flury, F. Uber das Aplysiengift. *Arch. exp. Pathol. Pharmokol.* 79, 250, 1915.

Fogh, R.H., Mabbutt, B.C., Kem, W.R. *et al.* Sequence-specific ^{1}H NMR assignments and secondary structure in the sea anemone polypeptide *Stichodactyla helianthus* neurotoxin I. *Biochemistry* 28, 1826, 1989.

Fogh, R.H., Kem, W.R. and Norton, R.S. Solution structure of neurotoxin I from the sea anemone *Stichodactyla helianthus*. *J. Biol. Chem.* 265(2), 13016, 1990.

Fontaine, A.R. The integumentary mucous secretions of the ophiuroid *Ophiocomina nigra*. *J. mar. Biol. Ass. U.K.* 44, 145, 1964.

Fraser, J.H. and Lyell, A. Dogger Bank itch. *The Lancet* 1, 61, 1963.

Franz, D.R., LeClaire, R.D. and Hernandez, D. Effects of saxitoxin on respiratory function in the awake guinea pig. *Toxicon* 27, 45, 1989.

Freeman, S.E. and Turner, R.J. A pharmacological study of the toxin of a cnidarian, *Chironex fleckeri* Southcott. *Br. J. Pharmacol.* 35, 510, 1969.

Freeman, S.E. and Turner, R.J. Maculotoxin, a potent toxin secreted by *Octopus maculosus* Hoyle. *Toxicol. appl. Pharmacol.* 16, 681, 1970.

Freeman, S.E., Turner, R.J. and Silva, S.R. The venom and venom apparatus of the marine gastropod *Conus striatus* Linné. *Toxicon* 12, 587, 1974.

Frey, D.G. The use of sea cucumbers in poisoning fishes. *Copeia* No. 2, 175, 1951.

Friess, S.L., Chanley, J.D., Hudak, W.V. *et al.* Interactions of the echinoderm toxin holothurin A and its desulfated derivative with the cat superior cervical ganglion preparation. *Toxin* 8, 211, 1970.

Friess, S.L. and Durant, R.C. Blockade phenomena at the mammalian neuromuscular synapse. Competition between reversible anticholinesterases and an irreversible toxin. *Toxicol. appl. Pharmacol.* 7, 373, 1965.

Friess, S.L., Durant, R.C., Chanley, J.D. *et al.* Role of the sulphate charge center in irreversible interaction of holothurin A with chemoreceptors. *Biochem. Pharmac.* 16, 617, 1967.

Friess, S.L., Durant, R.C. and Chanley, J.D. Further studies on biological actions of steroidal saponins produced by poisonous echinoderms. *Toxicon* 6, 81, 1968.

Friess, S.L., Standaert, F.G., Whitcomb, E.R. *et al.* Some pharmacologic properties of holothurin, an active neurotoxin from the sea cucumber. *J. Pharm. exp. Therap.* 126, 323, 1959.

Friess, S.L. Standaert, F.G., Whitcomb E.R. *et al.* Some pharmacologic properties of holothurin A, a glycosidic mixture from the sea cucumber. *Ann. N.Y. Acad. Sci.* 90, 893, 1960.

Fujiwara, T. [On the poisonous pedicellariae of *Toxopneustes pileolus* (Lamarck). *Annotationes zool. Jpn.* 15, 62, 1935.

Fukui, M., Murata, M., Inoue, A. *et al.* Occurrence of palytoxin in the trigger fish *Melichthys vidua*. *Toxicon* 25, 1121, 1987.

INDEX OF SCIENTIFIC NAMES